# A Pictorial Approach to Molecular Bonding

John G. Verkade

# A Pictorial Approach to Molecular Bonding

With 231 Illustrations

Springer-Verlag
New York Berlin Heidelberg
London Paris Tokyo

John G. Verkade
Department of Chemistry
Iowa State University of Science and Technology
Ames, Iowa 50011, U.S.A.

*541.224*
*V58p*
*141589*
*apm 1987*

Library of Congress Cataloging in Publication Data
Verkade, John G., 1935–
  A pictorial approach to molecular bonding.
  Includes bibliographical references and index.
  1. Molecular bonds. I. Title.
QD461.V45   1986     541.2'24     85-30272

Typeset by Asco Trade Typesetting Ltd., Hong Kong.
Printed and bound by R. R. Donnelley and Sons, Harrisonburg, Virginia.
Printed in the United States of America.

9 8 7 6 5 4 3 2 1

ISBN 0-387-96271-9 Springer-Verlag New York Berlin Heidelberg
ISBN 3-540-96271-9 Springer-Verlag Berlin Heidelberg New York

Aan mijn familie

# Preface

With the development of accurate molecular calculations in recent years, useful predictions of molecular electronic properties are currently being made. It is therefore becoming increasingly important for the non-theoretically oriented chemist to appreciate the underlying principles governing molecular orbital formation and to distinguish them from the quantitative details associated with particular molecules. It seems highly desirable then that the non-theoretician be able to deduce results of general validity without esoteric mathematics. In this context, pictorial reasoning is particularly useful. Such an approach is virtually indispensable if bonding concepts are to be taught to chemistry students early in their careers.

Undergraduate chemistry majors typically find it difficult to formulate molecular orbital schemes, especially delocalized ones, for molecules more complicated than diatomics. The major reason for this regrettable situation is the general impracticability of teaching group theory before students take organic and inorganic courses, wherein the applications of these concepts are most beneficial. Consequently many students graduate with the misconception that the ground rules governing bonding in molecules such as $NH_3$ are somehow different from those which apply to aromatic systems such as $C_6H_6$. Conversely, seniors and many graduate students are usually only vaguely, if at all, aware that sigma bonding (like extended pi bonding) can profitably be described in a delocalized manner when discussing the UV-photoelectron spectrum of $CH_4$, for example. Moreover, many graduate students who have had group theory find it difficult to visualize pictorially the linear combinations of AOs which make up MOs and to picture the relative movements of the atoms in the normal vibrational modes of even very symmetrical polyatomic systems.

In 1968, Professor Klaus Ruedenberg and the author became aware of this dilemma, and to remedy the situation we jointly developed a new course designed to teach the basic elements of chemical bonding to undergraduates. In order to do justice to the point of view of the theoretical as well as the experimental chemist, we team-taught the course for several years and, as a result of many hours of teaching and many more hours of discussion, a set of printed class notes came into existence. Using the nodal symmetries of atomic orbitals, Klaus outlined during one of these warmly acknowledged conversations how the intuitive reasoning associated with the "united atom model," originally introduced by R. S. Mulliken in connection with correlation diagrams, can provide a tool for students to learn how to construct MOs in very simple systems of high symmetry. Further reflection along these lines led the author to determine with simple sketches whether the nodal properties of atomic orbitals placed at the center of more complicated and less symmetrical molecules could be effectively used as a device to generate their MOs as LCAOs. This "generator orbital" method appeared to be widely applicable and, after Klaus justified its generality on group theoretical grounds in 1973, we used this approach successfully ever since in a course on bonding taught mainly to undergraduate majors and interested graduate students. In 1975, Professor David K. Hoffman pointed out that the generator orbital approach can also be used for a pictorial deduction of localized MOs and of normal vibrational modes in molecules. Dave's contributions of these ideas are also most warmly acknowledged and their development is included here.

Although the generator orbital concept can be presented in a mathematical and group theoretical framework, the concept and its applications lend themselves exceedingly well to a pictorial approach. The only prerequisites are high school level chemistry, geometry, physics, and trigonometry. Although integrals, vectors, and matrices are *briefly* touched upon, a working knowledge of these subjects is not necessary for understanding the generator orbital concept and its applications.

Some of the results of the fruitful collaborations with Klaus and David were published in article form.[a] We believe that more widespread pedagogical benefit can be realized by developing a textbook containing the ideas and applications contained in these papers, as well as in the class notes which by 1977 had become quite voluminous. After the appearance of the articles in that year, the three of us completed a preliminary manuscript for such a book and made several serious attempts to bring it into publishable form. Because of the press of other commitments as well as philosophical differences (stemming from our respective scientific backgrounds) concerning the manner of presentation of the material, it devolved on the author, by mutual agreement,

---

[a] See D. K. Hoffman, K. Ruedenberg, and J. G. Verkade, "A Novel Pictorial Approach to Teaching MO Bonding Concepts in Polyatomic Molecules," J. Chem. Ed., *54*, 590 (1977); and D. K. Hoffman, K. Ruedenberg, and J. G. Verkade, "Molecular Orbital Bonding Concepts in Polyatomic Molecules: A Novel Pictorial Approach," Structure and Bonding, *33*, 59 (1977).

to finish the project. The author, therefore, takes responsibility for errors and ambiguities which will undoubtedly be found, and it is hoped that these will be brought to his attention. The contents of essentially all of the first three chapters were adapted from more extensive class notes prepared by Klaus, and substantial portions of Chapters 4, 5, and 6 were developed in more extensive form by Klaus and David. David also made many helpful suggestions for the remaining chapters.

Many teachers of inorganic chemistry are currently seeking ways to reintroduce more descriptive chemistry at both the undergraduate and graduate levels. Part of the motivation in writing this book was to provide students with some simple tools for rationalizing the bonding in a very wide variety of molecules in a highly unified theoretically sound manner. By utilizing the generator orbital approach, the total time spent in teaching bonding concepts in inorganic, organic, and physical chemistry courses can actually be significantly reduced.

My loving thanks go to my wife, Sue, whose support and patience were truly wonderful in helping me finish this project. I also thank Mrs. Joyce Gilbert and Mrs. Peggy Biskner for their excellent deciphering and typing skills in bringing my handwritten manuscript into readable form. My thanks also go to Professor Walter Struve and the members of my 1984 and 1985 Structure and Bonding classes for reading the final manuscript and making many helpful suggestions.

Ames, Iowa                                                                                  John Verkade
July 1986

# Contents

# CHAPTER 1

# The Orbital Picture for Bound Electrons

The primary purpose of this book is to develop for the non-theoretically oriented chemist a pictorial approach to molecular orbitals. To accomplish this goal we consider in this chapter and Chapter 2 the fundamental properties of electrons and orbitals which are required for the introduction of "generator orbitals." Generator orbitals (which will be defined later) constitute the tool with which the three-dimensional visualization of molecular orbitals is achieved in subsequent chapters. We will see that the vibrational modes of molecules are also easily visualized with this approach. After we apply the generator orbital concept in Chapter 3 to bonding and to vibrational motion in diatomics, we will apply this concept for the same purposes to more complex polyatomic molecules possessing a variety of geometrical shapes.

## 1.1. Traveling and Standing Waves

Since we will soon appreciate that electrons can behave as either traveling or standing waves, it is important to understand some properties of waves. When a guitar string is plucked, the magnitude of the string displacement is called its *amplitude*. The louder the instrument is played, the greater the amplitude. It is interesting, however, that the amplitude of a plucked string varies over its length, and it also changes with time. Thus *the amplitude for such a one-dimensional wave is a function of both position and time: $f(x, t)$*. As we will see shortly, the standing wave pattern in a plucked string fixed at both ends is composed of two traveling waves. In a traveling wave, an observer sees a sinusoidal pattern moving with constant velocity along the $x$ direction (i.e., the length) of the string. The amplitude or displacement function for the wave is given by Equation 1-1; alternatively the sine function may be substituted by

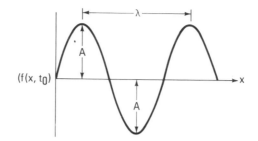

Figure 1-1. A "snapshot" of a sinusoidal wave traveling in the $x$ direction with wavelength $\lambda$.

the cosine function:

$$f(x, t) = A \sin 2\pi \left( \frac{x}{\lambda} - \frac{t}{\tau} \right). \tag{1-1}$$

Here $x$ measures the distance along the string, $t$ is the time, $A$ is the maximum amplitude, and $\lambda$ and $\tau$ are constants having the same units as $x$ and $t$, respectively. The dimension of $f$ is that of $A$, which in the case of the string would be in units of length.

To see the meaning of $\lambda$, suppose the wave is frozen at a fixed time $t = t_0$. The amplitude is now a sinusoidal function of space only (see Figure 1-1). *The distance at which the entire pattern repeats itself is called the wavelength, $\lambda$.* The reciprocal of $\lambda$ is called the *wave number*, $\bar{\nu}$, and it represents the number of full waves per centimeter.

In order to appreciate the meaning of $\tau$, suppose an observer views the wave at a fixed point $x = x_0$. The observer will then see an amplitude which oscillates from a positive value to an equal but negative value with time (Figure 1-2). The time it takes for the amplitude to return to its initial value is called the *period*, $\tau$. The number of times this phenomenon occurs per second

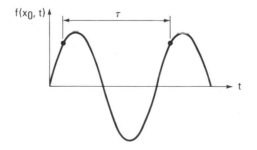

Figure 1-2. The behavior of the amplitude of a traveling wave as viewed at a fixed point $x = x_0$, where $\tau$ is the period.

is given by $1/\tau$, and this quantity is called the *frequency*, $\nu$. The speed $v$ of an advancing crest in a traveling wave is given by $v = \lambda \nu$.

The surface of a wave coming in on a beach is a two-dimensional traveling wave (Figure 1-3) since it has troughs and crests possessing length along $y$. The amplitude function is again given by Equation 1-1 and is a constant along the $y$ direction. To visualize a three-dimensional traveling wave, consider what happens when a sound wave is produced. The air, which was of constant density before the sound, now possesses regions of higher and lower density which are depicted in Figure 1-4, wherein only the $x$ and $z$ dimensions are shown for clarity. The density in any given $yz$ plane is constant. The propagation of such a plane wave is characterized by positive and negative deviations from the average density of air. The magnitude of these deviations is the amplitude.

To understand *standing* waves, it helps to examine the traveling waves we have considered for points, lines, or planes where the amplitude is zero and where it is maximum. Zero-amplitude regions are called *nodes*, and *antinodes* refer to places of maximum amplitude. In a traveling wave, nodes and antinodes move at the same velocity as the whole pattern. In a standing wave, the nodes and antinodes are fixed in space. The amplitude of a standing wave is given by any one of the expressions in Equation 1-2:

$$A\left(\sin 2\pi \frac{t}{\tau}\right)\left(\sin 2\pi \frac{x}{\lambda}\right)$$

$$A\left(\sin 2\pi \frac{t}{\tau}\right)\left(\cos 2\pi \frac{x}{\lambda}\right)$$

$$A\left(\cos 2\pi \frac{t}{\tau}\right)\left(\sin 2\pi \frac{x}{\lambda}\right) \tag{1-2}$$

$$A\left(\cos 2\pi \frac{t}{\tau}\right)\left(\cos 2\pi \frac{x}{\lambda}\right).$$

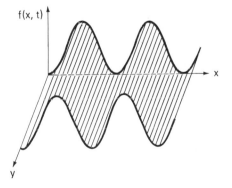

Figure 1-3. A two-dimensional traveling wave represented by the surface of an ocean.
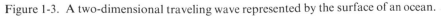

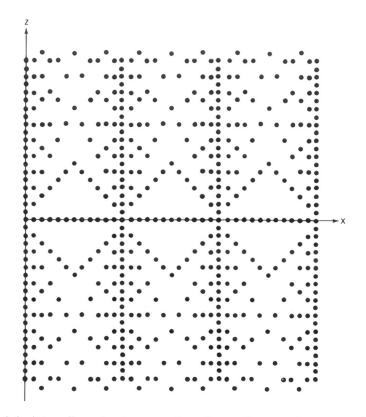

Figure 1-4. A two-dimensional cross section of a traveling sound wave caused by the *yz* surface vibrating back and forth across $x = 0$, thereby causing compressions and rarefactions among the air molecules to move along the *x* direction.

In the first expression the amplitude vanishes for $x = 0$ and at every half-wavelength [i.e., $x = (1/2)\lambda$, $\lambda$, $(3/2)\lambda$, $2\lambda$, etc.]. The antinodes are seen from this expression to appear halfway between the nodes. In the remaining expressions, nodes and antinodes appear at different $x$ values. In all the expressions, however, the nodes and antinodes are at fixed values of $x$ and are independent of time. There is an important difference between nodes and antinodes in a standing wave. Nodes are time-independent points, lines, or planes while antinodes oscillate with time. In a one-dimensional wave the amplitude points oscillate between positive and negative values of $z$ (Figure 1-5) and in a two-dimensional wave the amplitude line does the same. In a three-dimensional plane wave, a density oscillates in the *yz* plane.

Standing waves occur when *boundary conditions* are imposed. Clamps at both ends of a vibrating string impose the important boundary condition that there be nodes at these endpoint positions. The wave, therefore, *must* be a standing wave and in fact the length ($L$) of the domain between the endpoints

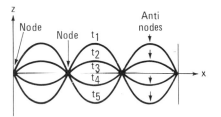

Figure 1-5. A standing wave bounded at the extreme ends, showing the time-independent nodes and time-dependent antinodes.

must be a multiple of the length between any *two* nodes ($\frac{1}{2}\lambda$). Thus the only wavelengths permitted are $\lambda_n = 2L/n$, with $n = 1, 2, 3, 4$, etc. In other words, $\lambda$ is *quantized* to particular values and the $\lambda_n$ are the *characteristic* or "*eigen*" values of the wavelength.

It is an interesting observation that two identical traveling waves propagating in opposite directions form a standing wave when superposed. Thus, for example, even though neither one of two superposed traveling one-dimensional waves contains points at which the amplitude vanishes *at all times*, such points (the nodes) do exist in the standing wave.

## 1.2. Wave Energy and Interference

*All waves contain energy*. Such energy can be kinetic and potential energy of matter (as for water or sound waves) or it can be electromagnetic (as in light waves). In all cases, the *amount of wave energy per unit volume* (i.e., the energy density or intensity) *is proportional to the square of the amplitude* (e.g., $f^2$ in Equation 1-1). In a traveling wave, energy flows along the direction of propagation. In a standing wave the energy is *localized* in the regions between the nodes and the nodes always have zero energy density.

If two or more sound waves having different frequencies encounter each other, their amplitudes superpose to form a new wave $f(x, y, z, t) = f_1(x, y, z, t) + f_2(x, y, z, t) + \cdots + f_n(x, y, z, t)$. Because the amplitudes of the constituent waves enhance one another at some points and cancel each other at others, a new set of oscillations is created. This has important consequences for the composite wave. Thus the energy density at most points in the new wave [i.e., $f^2 = (f_1 + f_2 + f_3 + \cdots + f_n)^2$] is not the same as the sum of the energy densities of the constituent waves [i.e., $f_1^2 + f_2^2 + \cdots + f_n^2$]. The difference of these two sums plotted along the direction of propagation gives an *interference* pattern. Such interference of two frequencies (Figure 1-6a) gives rise to the "beat" (Figure 1-6b) of fluctuating sound intensity heard when two identical or different instruments are playing slightly different pitches (frequencies) at the same intensity.

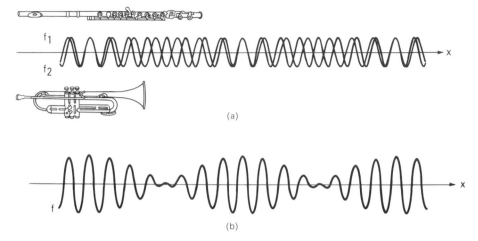

(a)

(b)

Figure 1-6. Superposition of two traveling waves of different frequency (a) to give a composite traveling wave (b) having "beats" of intensity. The closer the two frequencies are to each other, the closer together the "beats" are heard.


## 1.3. Electron Orbitals

*Under many conditions, electrons behave as waves while under others they possess the characteristics of particles.* This should not be disconcerting. The fundamental mathematical theory actually encompasses both the classical (particle) and the quantum mechanical (wave) descriptions harmoniously. It is only because of our limited ability of visualization that these approaches appear to be so different. Free electrons (i.e., in the absence of external forces and thus moving with constant velocity in a straight line) are found to act like three-dimensional traveling plane waves. In certain types of planar molecules such as butadiene ($CH_2$=CH—CH=$CH_2$) electrons behave like plane waves enclosed in a box. In a molecule in which double bonds alternate with single bonds (i.e., *conjugated* $\pi$ bond systems), the $\pi$ electrons (of which there are four in butadiene) move across *all the carbon atoms and also spend part of their time between the central two carbons*, even though the way we normally write the formula does not indicate this. In many respects the behavior and properties of these $\pi$ electrons can be described in terms of a free electron wave constrained to move along the carbon–carbon bond skeleton. Although we will not detail such a description here, these boundary conditions (as in the case of the string clamped at both ends) lead to quantized wavelengths $\lambda_k = 2L/k$, where $L$ is related to the length of the carbon skeleton and $k = 1$, 2, 3, etc. The kinetic energy in ergs of a free electron (which is the only kind of energy it has, even when constrained to remain in a confined space) is given by Equation 1-3 (which we do not derive), where $m$ = the mass of the electron and $h$ is Planck's constant ($6.6 \times 10^{-27}$ erg-sec). This energy quantization is

$$\varepsilon_k = \left(\frac{h^2}{8mL^2}\right)k^2 \tag{1-3}$$

characteristic of *all* electrons in atoms as well as molecules. Such electrons are all restricted to certain regions of space by electrostatic nuclear attractions and are said to be *bound* electrons. *All bound electrons (atomic and molecular) have wave character and possess quantized energy levels.* Although unbound electrons also have wave properties, their energies are not quantized. Quantization is the result of imposing boundaries on a wave. The energy levels of bound electron waves are spaced differently from one atom or molecule to another.

What does an amplitude function of an electron wave signify? For a string it is displacement of a point along $z$, for a water wave it is a similar displacement of a line parallel to $y$, and for a sound wave it is a deviation of air density from the average. For electron waves the amplitude function is related to the *distribution of the electron density* and is denoted by $\psi(x, y, z, t)$. How we reconcile the ideas of electrons as particles and as clouds having density will be made more clear shortly. For a free electron, $\psi$ has the mathematical features described earlier for traveling waves, but for a bound electron (i.e., a localized electron wave with boundary conditions on its length) the properties of $\psi$ are those of standing waves. In accord with Equation 1-2, a *three-dimensional* standing electron wave is a product of a space function $\phi_n$ and a time function $f_n$, i.e.,

$$\psi_n(x, y, z, t) = \phi_n(x, y, z)f_n(t). \tag{1-4}$$

For a one-dimensional standing wave this space factor has the sinusoidal form

$$\phi_n(x) = A \sin\frac{2\pi x}{\lambda_n}. \tag{1-5}$$

It contains the quantized wavelength $\lambda_n$ discussed earlier which, in turn, is related to the orbital energy $\varepsilon_n$ given by Equation 1-3. The *space* amplitudes $\phi_n$ associated with the first four $\varepsilon_n$ values (i.e., $n = 1, 2, 3, 4$) are shown in Figure 1-7. The time factors of Equation 1-4 for different bound electrons are given by $f_n = \exp(-i\varepsilon_n t/\hbar)$. Their functional form is thus the same for all states and it is only the value of the energy $\varepsilon_n$ which differentiates between different states and different systems. *However, the spatial amplitudes for bound electrons differ markedly, and it is these functions that characterize the individual natures of various states.* In general (i.e., for electrons under the influence of forces) these space amplitudes are three-dimensional in nature $[\phi_n(x, y, z)]$ and their shapes are not sinusoidal. One important feature of the electron waves shown in Figure 1-7 which carries over into *all* standing electron waves is that they too possess *nodes*, *positive lobes* (amplitudes), and *negative lobes*.

Single-electron standing waves are called *orbitals* and this term applies to the total wave function $\psi_n(x, y, z, t)$ as well as to only its spatial amplitude $\phi_n(x, y, z)$. An electron described by $\psi_n$ is said to occupy the orbital $\psi_n$ or $\phi_n$.

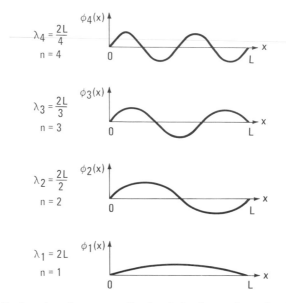

Figure 1-7. A series of space amplitudes $\phi_n$ for four values of $\varepsilon_n$ ($n = 1$ to $4$).

## 1.4. Normalization and Orthogonality

These formidable words denote rather simple concepts which can be developed after we consider how the electron density is related to the electron wave function. Earlier it was stated that the amplitude function $\psi$ "describes the distribution of an electron." How can this be if $\psi$ can have negative values? The answer is that *the fraction of the electron cloud which is found at the time t in the infinitesimally small volume element $dV = dxdydz$ enclosing the point $x, y, z$ is given by $|\psi(x, y, z, t)|^2 dV$, and this quantity is always positive.* (Recall that the letter $d$ in $dx$, etc., simply means an infinitesimally small increment.) This means that at time $t$ the volume element $dV$ makes a contribution of $m|\psi|^2 dV$ to the total electron mass and a contribution $-e|\psi|^2 dV$ to the total electron charge. The quantity $|\psi|^2$ is called the *orbital density*. More rigorously it describes the *probability* of finding the electron in $dV$.

Because the time factor in Equation 1-4 is of the form $\exp(i\alpha)$ with $\alpha$ being real, one finds $|\psi|^2 = \phi^2 |f|^2 = \phi^2(x, y, z)$. This shows that bound orbital densities do not change with time and that they can be calculated. Since $\phi^2$ tells us the essential characteristics of a given system, we will deal mainly with the *spatial orbitals* $\phi(x, y, z)$ rather than total functions $\psi(x, y, z, t)$.

Since $\phi^2$ describes one whole electron in the entire space and $\phi^2 dV$ represents the fraction of this electron in the volume $dV$, we can add up all the $\phi^2 dV$ in the space and say that it is equal to one electron. The mathematical procedure for such addition is called *integration* and its notation is shown in Equation 1-6.

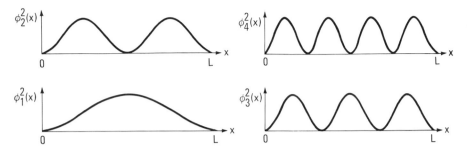

Figure 1-8. The $\phi_n$ of Figure 1–7 are squared to show their probability densities $\phi_n{}^2(x)$.

$$\int dV\,\phi^2(x, y, z) = 1 \qquad (1\text{-}6)$$

This requirement is called the *normalization* of the orbital $\phi$ and it says that the single electron is contained in the space defined by adding up all the infinitesimally small volume elements $dV$ which contain electron density.

In the case of a one-dimensional standing electron wave (Figure 1-7) $\phi_n{}^2(x)dx$ is the fraction of an electron in $\phi_n$ which lies between $x$ and $(x + dx)$. The probability densities $\phi_n{}^2(x)$ have the forms shown in Figure 1-8.

To help us understand orbital occupancies of ground and excited states of a system, it is very helpful to relate the geometrical shapes of different orbitals to their energies. Much information of this kind is contained in the nodes and antinodes (lobes) of orbitals. For example for our one-dimensional electron wave, the energies of the orbitals $\phi_n$, as given by Equation 1-3 rise monotonically with $n$. Note that the number of nodes *inside* the box is given by $n - 1$. To help us generalize this observation to other systems, the concept of *orbital orthogonality* is important. To appreciate this concept consider what happens when we generalize the normalization integral (Equation 1-6) to *two* different spatial orbitals $\phi$ and $\psi$ as shown in Equation 1-7. Since the time factor is no longer necessary for us, $\psi$ will be used to denote another spatial orbital. The

$$\int dV\,\phi(x, y, z)\psi(x, y, z) = \langle\phi|\psi\rangle \qquad (1\text{-}7)$$

integral in Equation 1-7 (for which we will use the shorthand notation $\langle\phi|\psi\rangle$) is called the *overlap integral* between $\phi$ and $\psi$. This integral is zero for two normalized orbitals on the same atom. Orbitals for which the overlap integral is zero are said to be *orthogonal* to each other. Strictly speaking this applies only to nondegenerate orbitals. Degenerate orbitals such as $2px$, $2py$, $2pz$ are indeed orthogonal to one another, but as we shall see later, linear combinations of these orbitals need not be. Orthogonality does not imply that the two orbitals are at right angles to each other in space. It means only that their overlap is zero.

The relation between orbital orthogonality and orbital nodal character can be seen by examining the one-dimensional electron wave. For example, are $\phi_1$ and $\phi_2$ orthogonal, i.e., is $\langle \phi_1 | \phi_2 \rangle = 0$? The product of the two orbitals is plotted in Figure 1-9. If we integrate over the distance 0 to $L$, the contributions of the positive and negative lobes cancel by symmetry. Thus the integral $\langle \phi_1 | \phi_2 \rangle$ is equal to 0 and the orbitals are orthogonal. *The pivotal point here is that one of the orbitals ($\phi_2$) has a node where the other ($\phi_1$) does not.* Similar reasoning applies to three-dimensional orbitals. Each orbital of an atom or molecule has a characteristic nodal structure which renders all the nondegenerate orbitals orthogonal to each other in the same way that $\phi_1$ and $\phi_2$ in Figure 1-9 are. As a general rule the number of nodes of an orbital in a given system increases with the orbital energy.

It is important to note that observations on mass or charge density of the electron tell us only about $|\psi|^2$ but not directly about $\psi$. In fact, there is no observable property of the electron which allows us to observe $\psi$ directly. To calculate any electronic property, however, we must in general have the amplitude $\psi$. Thus an important area of research is the development of more accurate wave functions which describe electron behavior in atoms and molecules. It should be noted that the calculation of any observable property always involves the wave function $\psi$ multiplied either by itself or by some derivative of $\psi$. Consequently, the wave function $\psi$ yields the same value for any property as the wave function $-\psi$. The sign of $\psi$ is therefore arbitrary. In other words, $\psi$ and $-\psi$ are two equivalent alternative mathematical descriptions of the same physical orbital.

## 1.5. Systems with Many Electrons

In atoms and molecules with many electrons, electrons occupy various orbitals, and quantum states of the total system differ by their occupation patterns. The possible occupation patterns are restricted by the *Pauli exclusion principle*, which specifies that each orbital cannot be occupied by more than two electrons. Moreover, if any orbital contains two electrons, they must have opposite spin rotations. The concept of an electron as a particle spinning on its axis seems incongruous when we are describing the wave nature of an electron in a confined space. Recall that particle and wave descriptions

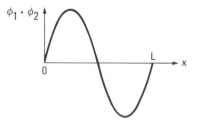

Figure 1-9. Plot of $\phi_1 \cdot \phi_2$ showing the total area under the curve to be zero.

are only mathematical models which help us understand submicroscopic phenomena.

*The total energy of a system of electrons in a molecule depends upon the energies of the quantized orbitals which are occupied by electrons.* This overall energy can often be approximated by the sum of the energies of all the electrons. In butadiene, for example, we can sketch the *orbital energy level diagram* for the $\pi$ system as shown in Figure 1-10a. Here $E_0$ represents the lowest energy or *ground state* of the $\pi$ system with two electrons in $k = 1$ and two electrons in $k = 2$ (Equation 1-3). Two other possible occupation patterns are shown in Figure 1-10b and c. These patterns are called *excited states* since their total energies are larger than $E_0$. Thus $E_0 = 2\varepsilon_1 + 2\varepsilon_2$ and $E_1 = 2\varepsilon_1 + \varepsilon_2 + \varepsilon_3$. The *total* energies of the ground and excited states relative to $\varepsilon_1$, and $\varepsilon_2$, and $\varepsilon_3$ are shown in Figure 1-10d. Excited states can be achieved by pumping energy into the ground state. The frequency of electromagnetic radiation required for such *absorption* of energy is given, for example, by $\Delta E = E_1 - E_0 = h\nu$ in Figure 1-11. This frequency can be radiated by the molecule when the excited state returns to the ground state (*emission*). Thus it is seen that molecules have *quantized energy states* as a consequence of electrons occupying *quantized energy levels*. The family of all permitted electronic transitions constitutes the *electronic spectrum* of a molecule.

In many-electron atoms and molecules, the total energy is a complicated function of nuclear charges and interelectronic repulsions. All formulas describing properties of many-electron systems, however, have one feature in common; namely, that by observing the properties of the entire system it is not possible to tell which electron occupies which orbital. The electrons are said to be *indistinguishable*.

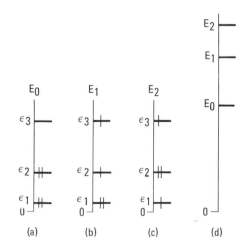

Figure 1-10. Orbital energy level diagrams for a ground (a) and two excited [(b), (c)] states and an energy level diagram (d) for the states represented in (a)–(c).

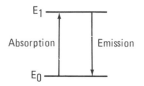

Figure 1-11. Energy level diagram depicting absorption of energy required for excitation from $E_0$ to $E_1$ and the emission of an equal amount of energy to descend from $E_1$ to $E_0$.

## 1.6. Flexibility of Orbital Sets

An interesting feature of many-electron systems, which is related to the indistinguishability of electrons, is the fact that the orbitals which are occupied in a particular state of a system are not unique. That is to say, one can often make certain alterations in the individual occupied orbitals without changing the values of any of the overall electronic properties of the system. This remarkable concept is frequently very useful in understanding the bonding in molecules.

To keep matters simple, suppose we have the three orbitals $\phi_1$, $\phi_2$, $\phi_3$ which are known, and another set $\psi_1$, $\psi_2$, $\psi_3$ which we constructed from $\phi_1$, $\phi_2$, $\phi_3$ by Equation 1-8 wherein the subscripts in the $T_{jk}$ coefficients are merely labeling numbers. We can write a shorthand version of Equation 1-8 called a

$$\psi_1 = T_{11}\phi_1 + T_{12}\phi_2 + T_{13}\phi_3$$

$$\psi_2 = T_{21}\phi_1 + T_{22}\phi_2 + T_{23}\phi_3 \qquad (1\text{-}8)$$

$$\psi_3 = T_{31}\phi_1 + T_{32}\phi_2 + T_{33}\phi_3$$

*matrix* (Equation 1-9), in which the first index on the coefficients $T_{jk}$ denotes

|          | $\phi_1$ | $\phi_2$ | $\phi_3$ |
|----------|----------|----------|----------|
| $\psi_1$ | $T_{11}$ | $T_{12}$ | $T_{13}$ |
| $\psi_2$ | $T_{21}$ | $T_{22}$ | $T_{23}$ |
| $\psi_3$ | $T_{31}$ | $T_{32}$ | $T_{33}$ |

$$(1\text{-}9)$$

the *row* and the *second* denotes the *column* of the matrix. Each of Equations 1-8 is obtained by multiplying the $\phi_n$ by the coefficient below them in the row corresponding to the $\psi_n$ and adding. Because orbitals must be normalized (Equation 1-6) and mutually orthogonal (i.e., zero overlap, Equation 1-7), the *two sets* of integrals $\langle \phi_j | \phi_k \rangle$ and $\langle \psi_j | \psi_k \rangle$ must both be 1 for $j = k$ and 0 for $j \neq k$. We now demonstrate that these conditions lead to important restrictions on the values the coefficients $T_{jk}$ can take.

Selecting $\psi_1$ and $\psi_2$ as an example, we substitute the expansions for these orbitals given in Equation 1-8 into the overlap integral $\langle \psi_1 | \psi_2 \rangle$ in Equation 1-10:

$$\langle \psi_1 | \psi_2 \rangle = \langle T_{11}\phi_1 + T_{12}\phi_2 + T_{13}\phi_3 | T_{21}\phi_1 + T_{22}\phi_2 + T_{23}\phi_3 \rangle. \quad (1\text{-}10)$$

Carrying out the multiplication indicated in Equation 1-10, we obtain Equation 1-11:

$$\begin{aligned}
\langle \psi_1 | \psi_2 \rangle = {} & T_{11}T_{21}\langle \phi_1 | \phi_1 \rangle + T_{11}T_{22}\langle \phi_1 | \phi_2 \rangle + T_{11}T_{23}\langle \phi_1 | \phi_3 \rangle \\
& + T_{12}T_{21}\langle \phi_2 | \phi_1 \rangle + T_{12}T_{22}\langle \phi_2 | \phi_2 \rangle + T_{12}T_{23}\langle \phi_2 | \phi_3 \rangle \quad (1\text{-}11) \\
& + T_{13}T_{21}\langle \phi_3 | \phi_1 \rangle + T_{13}T_{22}\langle \phi_3 | \phi_2 \rangle + T_{13}T_{23}\langle \phi_3 | \phi_3 \rangle
\end{aligned}$$

which simplifies to Equation 1-12 when we apply the normalization and orthogonality criteria for $\phi_1$, $\phi_2$, and $\phi_3$:

$$\langle \psi_1 | \psi_2 \rangle = T_{11}T_{21} + T_{12}T_{22} + T_{13}T_{23}. \quad (1\text{-}12)$$

We also require, of course, that $\psi_1$ and $\psi_2$ be orthogonal and normalized, i.e.:

$$\langle \psi_1 | \psi_2 \rangle = T_{11}T_{21} + T_{12}T_{22} + T_{13}T_{23} = 0 \quad (1\text{-}13)$$

$$\langle \psi_1 | \psi_1 \rangle = T_{11}{}^2 + T_{12}{}^2 + T_{13}{}^2 = 1 \quad (1\text{-}14)$$

$$\langle \psi_2 | \psi_2 \rangle = T_{21}{}^2 + T_{22}{}^2 + T_{23}{}^2 = 1. \quad (1\text{-}15)$$

When functions are both orthogonal *and* normalized, we say they are *orthonormal*. The coefficients $T_{jk}$ must satisfy Equations 1-13 to 1-15 in order for $\psi_1$ and $\psi_2$ to be orthonormal *and* for $\phi_1$, $\phi_2$, and $\phi_3$ to be orthonormal. An example of such an array of numbers $T_{jk}$ which satisfies Equations 1-13, 14, 15 is shown in Equation 1-16. It is called an *orthogonal matrix*, and Equations 1-8 which generate the $\psi_m$ from the $\phi_n$ are called an *orthogonal transformation*.

$$\begin{bmatrix} T_{11} & T_{12} & T_{13} \\ T_{21} & T_{22} & T_{23} \\ T_{31} & T_{32} & T_{33} \end{bmatrix} = \begin{bmatrix} 1/6 & -5/6 & 10^{1/2}/6 \\ 10^{1/2}/6 & 10^{1/2}/6 & 4/6 \\ -5/6 & 1/6 & 10^{1/2}/6 \end{bmatrix} \quad (1\text{-}16)$$

Note that there was no need to be concerned about the explicit wave functions or the meaning of any of the implied integrations! The array of numbers in Equation 1-16 is only one example of an orthogonal array. Other orthogonal $3 \times 3$ arrays can be devised.

The two sets of orbitals $\psi_n$ and $\phi_n (n = 1, 2, 3)$ which we have been discussing are equivalent in that all observable electronic properties are independent of our choice of orbital sets. Suppose that $\psi_1$, $\psi_2$, and $\psi_3$ each contain the same number of electrons (say, two). The total density $\rho$ is then given by Equation 1-17, wherein the factor of 2 represents the presence of two electrons:

$$\rho = 2(\psi_1{}^2 + \psi_2{}^2 + \psi_3{}^2). \quad (1\text{-}17)$$

If we now use the relations of Equations 1-8 and the orthonormality conditions for $\psi_1$, $\psi_2$, $\psi_3$, Equation 1-18 can be obtained:

$$\rho = 2(\phi_1^2 + \phi_2^2 + \phi_3^2). \qquad (1\text{-}18)$$

Although the *individual* orbital densities of the $\phi_n$ are different from *individual* orbital densities of the $\psi_n$, the total density $\rho$ is independent of which orbital set we choose. The *individual orbital energies* are also different for the two sets but *the total energy is not dependent upon this choice*. In other words, *the state of the whole molecule or atom can be described by either set of orbitals.*

If a given molecular or atomic state can be described by a variety of orbital sets (as long as these sets are orthonormal) which set shall we choose? One set which is very useful is called the *canonical* set. *Canonical orbitals are the most fundamental type and they represent the various standing waves of a single electron as if it were moving alone in the total potential field generated by nuclei and electrons.* Consider the energies of a ground and first excited state given by Equation 1-19:

$$\begin{aligned} E_0 &= 2\varepsilon_1 + 2\varepsilon_2 + 2\varepsilon_3 \\ E_1 &= 2\varepsilon_1 + 2\varepsilon_2 + \varepsilon_3 + \varepsilon_4, \end{aligned} \qquad (1\text{-}19)$$

and their difference, which is described by Equation 1-20:

$$E_1 - E_0 = \varepsilon_4 - \varepsilon_3. \qquad (1\text{-}20)$$

Since all the orbitals of $E_0$ are equally occupied, we can perform an orthogonal transformation to obtain an equivalent set $\psi_1', \psi_2', \psi_3'$ for which the energy can be written as

$$E_0 = 2\varepsilon_1' + 2\varepsilon_2' + 2\varepsilon_3'. \qquad (1\text{-}21)$$

The orbitals $\psi_1$ and $\psi_2$ in $E_1$ contain equal numbers of electrons (two) while $\psi_3$ and $\psi_4$ each contain one. Each of the doubly occupied orbitals can be replaced by an equivalent set $\psi_1'', \psi_2''$ and each of the singly occupied orbitals can be replaced by an equivalent set $\psi_3'', \psi_4''$ which is, of course, different from $\psi_1'', \psi_2''$. The state energy for $E_1$ is

$$E_1 = 2\varepsilon_1'' + 2\varepsilon_2'' + \varepsilon_3'' + \varepsilon_4''. \qquad (1\text{-}22)$$

The orbital energies $\varepsilon_j'$ and $\varepsilon_j''$ will not be the same and so no simplifications can be made in the energy calculation in Equation 1-23:

$$E_1 - E_0 = 2(\varepsilon_1'' - \varepsilon_1') + 2(\varepsilon_2'' - \varepsilon_2') + \varepsilon_3'' + \varepsilon_4'' - 2\varepsilon_3'. \qquad (1\text{-}23)$$

We can see then that the canonical orbitals (rather than other orthonormal equivalent orbital sets) furnish a better picture of the different energy states of a system and, in particular, of what occurs when molecules emit and absorb energy as a result of electronic transitions.

There are some problems, such as the analysis of chemical similarities between related molecules, for which orbital sets orthonormal to the canonical set are more useful. One such set is the *localized* orbital set in which orbitals are spatially separated insofar as possible. A good example is the $\pi$

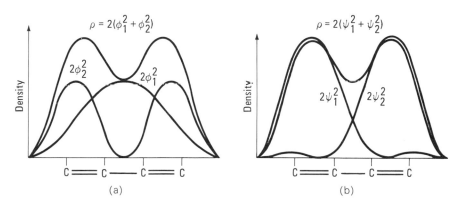

Figure 1-12. $\pi$ Electron density distributions in butadiene provided by canonical (a) and localized (b) orbitals.

electron *density distribution* for butadiene in Figure 1-12. Note that the total density is the same for both the canonical and the localized view, even though the individual orbital distributions in the two canonical orbitals $\phi_1$ and $\phi_2$ in (a) of Figure 1-12 differ from those of the localized orbitals $\psi_1$ and $\psi_2$ in (b) of this figure. In the localized approach each $\pi$ bond has associated with it an orbital of identical shape. To describe the $\pi$ orbitals in hexatriene ($CH_2$=CH—CH=CH—CH=$CH_2$) with localized orbitals, it turns out that all three $\pi$ bonds have $\pi$ orbitals which look similar to those in butadiene. Thus *localized orbitals are more enlightening when comparing ground states of similar molecules* whereas *canonical (more delocalized) orbitals are a more effective aid in comparing different states of the same molecule.*

## Summary

In this chapter the ideas of traveling and standing waves have been used to introduce the concept of electron orbitals. As with standing waves, the energies of bound electrons are quantized. The meaning and some of the utility of electron orbital normalization, orthogonality, and flexibility were also introduced.

### PROBLEMS

1. Two waves, $f(x, t) = \sin 2\pi[(x/\lambda) - (t/\tau)]$ and $f(x, t) = \sin 2\pi[(x/\lambda) + (t/\tau)]$, travel in opposite directions. Show by substituting numbers into the sum of these equations that their sum forms a plot of a standing wave.

2. Verify Equations 1-13–1-15.

3. Show that the numbers in Equation 1-16 satisfy Equations 1-13–1-15.

4. Obtain Equation 1-18 by the procedure suggested in the text.

5. The lowest two canonical orbitals in butadiene are the electron waves $\phi_1 = (2/5)^{1/2} \sin(\pi/5)x$ and $\phi_2 = (2/5)^{1/2} \sin(\pi/5)2x$, where the distance between two carbons is chosen as a unit of length, so that the distance $L$ within which the waves are contained is 5 units long (see Figure 1-12). The corresponding two localized orbitals are $\psi_1 = 2^{-1/2}(\phi_1 + \phi_2)$ and $\psi_2 = 2^{-1/2}(\phi_1 - \phi_2)$. (a) By substituting numbers, make a plot simultaneously showing $\psi_1(x)$, $\psi_2(x)$, $\psi_1^2(x)$, $\psi_2^2(x)$, and $[2\psi_1^2(x) + 2\psi_2^2(x)]$. (b) Do the same as in (a) for $\phi_1(x)$, $\phi_2(x)$, $\phi_1^2(x)$, $\phi_2^2(x)$, and $[2\phi_1^2(x) + 2\phi_2^2(x)]$. (c) Show from the plots that the members of each set are orthogonal. (d) Saying that $\langle \phi_1 | \phi_1 \rangle = 1$ when $\phi_1$ is normalized means that 1 is a unitless number. From the fact that $\phi_1^2 = (2/L)\sin(2\pi/L)x$, what must be the dimensions of $\phi_1^2$? Draw on your plot of $\phi_1^2$ a rectangle which has exactly the same area as that under your curve. Find the numerical value of this area and justify why it is unitless.

6. In hexatriene, the canonical orbitals are $\phi_n(x) = (2/7)^{1/2} \sin(\pi/7)nx$. In the ground state, the canonical orbitals $\phi_n$ and the localized orbitals $\psi_n$ are related by the matrix shown below. Show that this transformation is orthogonal and demonstrate by making plots that $\psi_1, \psi_2, \psi_3$ are quite localized in the double bond regions. In your plots, let $L = 7$.

|          | $\phi_1$     | $\phi_2$       | $\phi_3$      |
|----------|--------------|----------------|---------------|
| $\psi_1$ | 1/2          | $(1/2)^{1/2}$  | 1/2           |
| $\psi_2$ | $(1/2)^{1/2}$| 0              | $-(1/2)^{1/2}$|
| $\psi_3$ | 1/2          | $-(1/2)^{1/2}$ | 1/2           |

# CHAPTER 2

# Atomic Orbitals

One of our major goals is to appreciate the shapes and energies of the orbitals which electrons can occupy in atoms and molecules. In contrast to the orbitals in a one-dimensional box which we examined earlier, atomic and molecular orbitals are three-dimensional standing waves. We first address ourselves to the shapes of atomic *canonical* AOs (atomic orbitals) and their energies. Then we shall familiarize ourselves with alternative sets of AOs called *hybrids*. Canonical AOs have *identical shapes* in all states and in all atoms. They differ only in their *sizes*.

## 2.1. Shapes of Canonical AOs

The most important features of atomic orbitals are their three-dimensional *lobes*, where the amplitude $\psi$ is either positive or negative, and their nodes, where $\psi$ is zero. Because of the three-dimensional character of lobes, the nodes which separate lobes of opposite sign must be two-dimensional surfaces. The nodal surfaces of *canonical* AOs are *spheres*, *cones*, or *planes* and we will return to a discussion of these nodes shortly.

How big is an orbital? Because all atomic (and molecular) orbitals extend to infinity, their boundary at infinity can be viewed as a two-dimensional surface, namely, a sphere with $r = \infty$. We still can compare orbital sizes, however, because $\psi$ decays rapidly with $r$ in an approximately exponential fashion, and the rate of decrease in $\psi$ depends upon the orbital type. Thus $\psi(x, y, z) \approx K e^{-r/\alpha}$ (when $r \gg \alpha$) where $r = (x^2 + y^2 + z^2)^{1/2}$ is the distance of the electron from the atomic (or molecular) center, and $K$ and $\alpha$ are constants which are different for various orbitals. For example, let us consider the simplest of all atomic orbitals, the $1s$. Whenever we discuss the *spatial ampli-*

*tude* of an orbital we will parenthesize it [i.e., (1s) in the case at hand]. The spatial amplitude of the (1s) is described by

$$(1s) = \psi_{1s}(x, y, z) = (\pi\alpha^3)^{-1/2}e^{-r/\alpha} \tag{2-1}$$

where $\alpha$ is the ratio of the Bohr radius $a_0 = 0.529$ Å to the nuclear charge $Z$, and the factor $(\pi\alpha^3)^{-1/2}$ normalizes the wave function so that

$$\int dV(\psi_{1s})^2 = 4\pi \int_0^\infty dr\, r^2(\pi\alpha^{-3})^{-1}e^{-2(r/\alpha)}$$

$$= (4/\alpha^3)\int_0^\infty dr\, r^2 e^{-2(r/\alpha)} = 1. \tag{2-2}$$

For visualizing interatomic bonding, chemists prefer to think of orbitals as having sizes about as large as interatomic distances in molecules. Since orbital electron density falls off very rapidly with distance, disregarding only the outer 10% of an electron in any orbital brings the infinitely large orbital down in size so that the remaining 90% of its electron density is contained in a space which extends roughly to a neighboring atom in the molecule. The radii which contain 90% of an orbital's electron density vary with the orbital (i.e., they are proportional to $\alpha$), thus allowing comparisons of the "chemical reach" of different orbitals.

To describe the nodal surfaces of atomic orbitals, we will use the polar coordinate system in Figure 2-1 to describe a point $P$. As mentioned earlier, AOs have radial, planar, and conal nodes. *Radial nodes* are described by equations of the type $r =$ constant. These nodes are spheres with fixed radii and we denote the number of such spherical nodes by $N_r$. Here we adopt

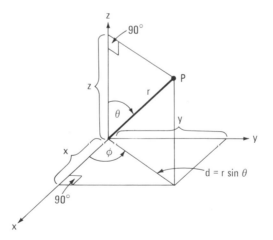

Figure 2-1. Polar coordinate system wherein $z = r\cos\theta$, $y = r\sin\theta\sin\phi$, $x = r\sin\theta\cos\phi$, $\tan\phi = y/x$, $\cos\theta = z/r$, and $r = (x^2 + y^2 + z^2)^{1/2}$. For $\theta$ and $\phi$ the relationships $0 \leq \theta \leq \pi$ and $0 \leq \phi < 2\pi$ hold, respectively.

the convention that the boundary at infinity is counted as one spherical node. *Planar angular nodes* are characterized by equations of the type $\phi$ = constant. These planes are perpendicular to the $xy$ plane, they contain the $z$ axis, and they form fixed angles with the $xz$ plane. The number of such nodes is given by $N_\phi$. *Conal angular nodes* are described by equations of the type $\theta$ = constant. The symmetry axis of these cones is $z$ and the angle between $z$ and a line on the conal surface is a fixed angle $\theta$. When $\theta < 90°$ the cone lies above the $xy$ plane and when $\theta > 90°$ the cone lies below this plane. When $\theta = 90°$, the cone lies in the $xy$ plane. This plane is a special case of a cone whose center axis is $z$. We will use the symbol $N_\theta$ to denote the number of conal nodes.

We now develop the concept that the nodal pattern as well as the shapes of the lobes are determined completely by the numbers $N_r$, $N_\theta$, and $N_\phi$. These numbers are related to the canonical AO quantum numbers $n$, $l$, and $m$ by Equations 2-3:

$$n = N_r + N_\theta + N_\phi = \text{total number of nodes}$$
$$\text{(including the one at infinity)}$$

$$l = N_\theta + N_\phi = \text{total number of conal angular and} \qquad (2\text{-}3)$$
$$\text{planar angular nodes}$$

$$m = N_\phi = \text{total number of planar angular notes.}$$

The $n$, $l$, and $m$ quantum numbers have the values $n = 1, 2, 3, 4, \ldots, l = 0, 1, 2, 3, \ldots, (n - 1)$ for a given $n$, and $m = 0, 1, 2, 3, \ldots, l$ for a given $l$. Thus if we know the values of $n$, $l$, $m$ we can calculate the number of nodal surfaces by Equations 2-4:

$$N_\phi = m$$
$$N_\theta = l - m \qquad (2\text{-}4)$$
$$N_r = n - l.$$

A common mode of designating AOs is by the *nlm system*. For historical reasons, letters have become associated with $l$ values, i.e., $s$ for $l = 0$, $p$ for $l = 1$, $d$ for $l = 2$, $f$ for $l = 3$, and then alphabetically starting with $g$ for succeeding $l$ values. Table 2-1 gives a breakdown of all orbitals from $(1s)$ to $(4f)$ in terms of the *nlm* symbolism. The meaning of $m$ and $m'$ and of the Cartesian labels will be discussed later.

Let us now consider what happens to an orbital shape as the number of nodes increases (i.e., as the energy rises). The ground state of any one-electron atom is the $(1s) = (100)$ orbital. Aside from the spherical boundary at infinity, it has no other node. As discussed earlier, the $(1s)$ distribution is spherically symmetrical and in fact this is true for all $s$ orbitals [i.e., $(1s)$, $(2s)$, $(3s)$, etc.]. Thus they all are functions of $r$ alone and as seen from Figure 2-2 they all possess $(n - 1)$ *finite* radial (spherical) nodes (plus an additional node at infinity). The intersections of the amplitudes with the $r$ axis correspond to the positions of the spherical nodes. Recall that the amplitude sign of any orbital

Table 2-1. Atomic Orbitals and their Quantum Numbers

| Orbital | $n$ | $l$ | $m$ | $N_\phi$ | $N_\theta$ | $N_r$ |
|---|---|---|---|---|---|---|
| $(1s0) = (1s)$ | 1 | 0 | 0 | 0 | 0 | 1 |
| $(2s0) = (2s)$ | 2 | 0 | 0 | 0 | 0 | 2 |
| $(2p0) = (2pz)$ | 2 | 1 | 0 | 0 | 1 | 1 |
| $(2p1') = (2px)$ | 2 | 1 | 1 | 1 | 0 | 1 |
| $(2p1'') = (2py)$ | 2 | 1 | 1 | 1 | 0 | 1 |
| $(3s0) = (3s)$ | 3 | 0 | 0 | 0 | 0 | 3 |
| $(2p0) = (3pz)$ | 3 | 1 | 0 | 0 | 1 | 2 |
| $(3p1') = (3px)$ | 3 | 1 | 1 | 1 | 0 | 2 |
| $(3p1'') = (3py)$ | 3 | 1 | 1 | 1 | 0 | 2 |
| $(3d0) = (3d(3z^2 - r^2)) = $ "$(3dz^2)$" | 3 | 2 | 0 | 0 | 2 | 1 |
| $(3d1') = (3dxz)$ | 3 | 2 | 1 | 1 | 1 | 1 |
| $(3d1'') = (3dyz)$ | 3 | 2 | 1 | 1 | 1 | 1 |
| $(3d2') = (3d(x^2 - y^2)) = $ "$(3dx^2 - y^2)$" | 3 | 2 | 2 | 2 | 0 | 1 |
| $(3d2'') = (3dxy)$ | 3 | 2 | 2 | 2 | 0 | 1 |
| $(4s0) = (4s)$ | 4 | 0 | 0 | 0 | 0 | 4 |
| $(4p0) = (4pz)$ | 4 | 1 | 0 | 0 | 1 | 3 |
| $(4p1') = (4px)$ | 4 | 1 | 1 | 1 | 0 | 3 |
| $(4p1'') = (4py)$ | 4 | 1 | 1 | 1 | 0 | 3 |
| $(4d0) = $ "$(4dz^2)$" | 4 | 2 | 0 | 0 | 2 | 2 |
| $(4d1') = (4dxz)$ | 4 | 2 | 1 | 1 | 1 | 2 |
| $(4d1'') = (4dyz)$ | 4 | 2 | 1 | 1 | 1 | 2 |
| $(4d2') = $ "$(4dx^2 - y^2)$" | 4 | 2 | 2 | 2 | 0 | 2 |
| $(4d2'') = (4dxy)$ | 4 | 2 | 2 | 2 | 0 | 2 |
| $(4f0) = (4f(5z^3 - 3r^2 z))$ | 4 | 3 | 0 | 0 | 3 | 1 |
| $(4f1') = (4f(5z^2 x - r^2 x))$ | 4 | 3 | 1 | 1 | 2 | 1 |
| $(4f1'') = (4f(5z^2 y - r^2 y))$ | 4 | 3 | 1 | 1 | 2 | 1 |
| $(4f2') = (4f(x^2 z - y^2 z))$ | 4 | 3 | 2 | 2 | 1 | 1 |
| $(4f2'') = (4f(xyz))$ | 4 | 3 | 2 | 2 | 1 | 1 |
| $(4f3') = (4f(x^3 - 3y^2 x))$ | 4 | 3 | 3 | 3 | 0 | 1 |
| $(4f3'') = (4f(3x^2 y - y^3))$ | 4 | 3 | 3 | 3 | 0 | 1 |

lobe is arbitary and we could just as well have shown plots obtained by inverting each of the ones in Figure 2-2 with respect to the $r$ axis.

The conal and planar angular nodes occur in certain patterns. When there is more than one conal node, they are always arranged symmetrically above and below the $xy$ plane. Thus whenever there is a cone at $\theta = $ some angle $\beta$ which is not $\pi/2$, then there is another cone which is its mirror image at $\theta = \pi - \beta$ on the other side of the $xy$ plane. If $\theta = \pi/2$, then the conal node in the $xy$ plane is its own mirror image. Whenever there is a cone with $\theta = \pi/2$, it means that the number of such nodes $(N_\theta)$ is odd. As shown in Figure 2-3, the first instance of this occurs at the orbital $(210) = (2pz)$.

Planar nodes where $\phi = $ constant are always arranged so that they divide the total angular range $0 \leq \phi < 2\pi$ into $2m$ equal parts, as shown in Figure 2-4. An additional property of these planar nodes is that for each value of $N_\phi$ (except zero) they have two independent arrangements in which the nodal planes of one orbital bisect the angles between the nodal planes of the other

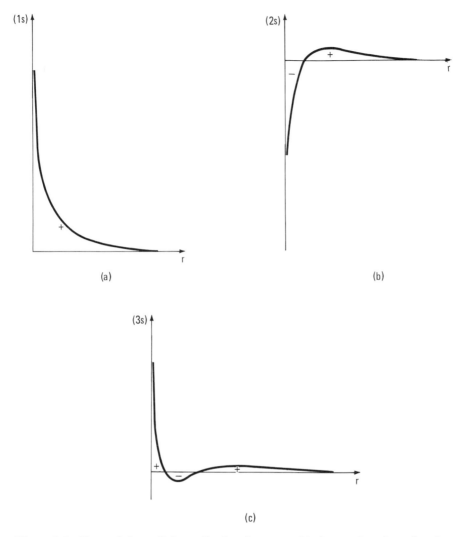

Figure 2-2. Plots of the radial amplitude of $s$-type orbitals as a function of $r$, the distance from the nucleus.

(Figure 2-5). The orbital which *does not* have a node in the $xz$ plane is designated by $m'$ and the orbital described which *has* a node in this plane is labeled $m''$. The nodal planes for the arrangements in Figure 2-5 are in Table 2-2. The equations of the nodal planes can be expressed in terms of Cartesian coordinates. Such notation, which is included as part of the expressions in Figure 2-5, is also frequently used in the shorthand designations for the orbitals instead of $m'$ or $m''$.

*The orbital lobes occur in the regions between the nodes.* These lobes and

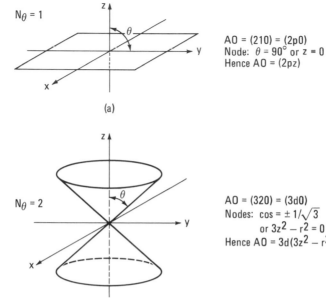

$N_\theta = 1$

AO = (210) = (2p0)
Node: $\theta = 90°$ or $z = 0$
Hence AO = (2pz)

(a)

$N_\theta = 2$

AO = (320) = (3d0)
Nodes: $\cos = \pm 1/\sqrt{3}$
    or $3z^2 - r^2 = 0$
Hence AO = $3d(3z^2 - r^2)$

(b)

$N_\theta = 3$

AO = (430) = (4f0)
Nodes: $\theta = 0$, $\cos \theta = \pm\sqrt{3/5}$
    or $z(5z^2 - 3r^2) = 0$
Hence AO = $4fz(5z^2 - r^2)$

(c)

Figure 2-3. Angular nodes for (a) $(2pz) = (2p0)$, (b) $(3dz^2) = (3d0)$, and (c) $(4f0)$.

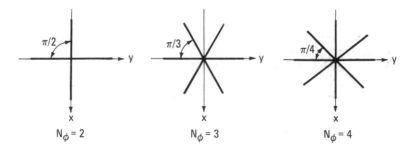

$\pi/2$                    $\pi/3$                    $\pi/4$

$N_\phi = 2$              $N_\phi = 3$              $N_\phi = 4$

Figure 2-4. Drawings showing how planar nodes divide the angular range $0 < \phi < 2\pi$ into $2m$ equal parts.

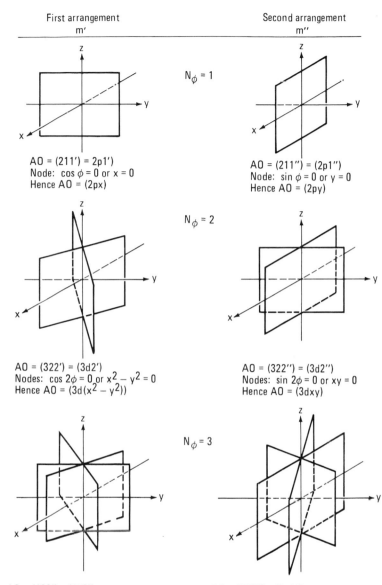

First arrangement
m'

Second arrangement
m''

$N_\phi = 1$

AO = (211') = 2p1')
Node: $\cos \phi = 0$ or $x = 0$
Hence AO = (2px)

AO = (211'') = (2p1'')
Node: $\sin \phi = 0$ or $y = 0$
Hence AO = (2py)

$N_\phi = 2$

AO = (322') = (3d2')
Nodes: $\cos 2\phi = 0$ or $x^2 - y^2 = 0$
Hence AO = $(3d(x^2 - y^2))$

AO = (322'') = (3d2'')
Nodes: $\sin 2\phi = 0$ or $xy = 0$
Hence AO = (3dxy)

$N_\phi = 3$

AO = (433') = (4f3')
Nodes: $\cos 3\phi = 0$ or $x(x^2 - 3y^2) = 0$
Hence AO = $(4fx(x^2 - 3y^2))$

AO = (433'') = (4d3'')
Nodes: $\sin 3\phi = 0$ or $y(y^2 - 3x^2) = 0$
Hence AO = $(4dy(y^2 - 3x^2))$

Figure 2-5. Drawing showing the two arrangements of vertical planes for $N_\phi = 1, 2, 3$.

Table 2-2. Nodal Planes and their Positions in the Atomic Orbital

| Number of Nodal Planes, $N_\phi = m$ | Arrangement with No Node in $xz$ Plane, $m'$ | Arrangement with a Node in $xz$ Plane, $m''$ |
|---|---|---|
| 1 | $\phi = 90°, x = 0$ | $\phi = 0°, y = 0$ |
| 2 | $\phi = 45°, x = y$ $\phi = -45°, x = -y$ | $\phi = 0°, y = 0$ $\phi = 90°, x = 0$ |
| 3 | $\phi = 30°, x = \sqrt{3}y$ $\phi = 90°, x = 0$ $\phi = -30°, \quad x = -\sqrt{3}y$ | $\phi = 0°, y = 0$ $\phi = 60°, y = \sqrt{3}x$ $\phi = -60°, y = \sqrt{3}x$ |

nodes are shown for the atomic canonical orbitals $(2s)$ through $(5g)$ in Figure 2-6. Although it is not apparent at this point why we need to concern ourselves with orbitals beyond $(4f)$, we will see in later chapters that an appreciation of the shapes of such orbitals is essential to the visualization of molecular orbitals and vibrational modes. We therefore now develop the simple rules for generating the shape of any canonical atomic orbital.

We begin with the general rule that the total number of nodes is $n$. Starting with the spherical orbital $(n00) = (ns)$, the nonspherical orbitals in Figure 2-6 can be visualized as will now be shown for $n = 3$. The (3s) = (300) AO has two finite spherical nodes. Converting one spherical node to a conal node gives us a $(3p0) = (3pz)$ orbital. Here the conal node must be in the $xy$ plane since we have only one such node. Converting the spherical node to a planar node leads to the $(3py)$ and $(3px)$ AOs, as shown in Figure 2-6. Similarly, converting both finite spherical nodes to two conal nodes gives us a $(3d0) = (3dz^2)$ AO (Figure 2-6) and, changing the two conal nodes one at a time to planar nodes generates the remaining four $(3d)$ orbitals (Figure 2-6). *The radial node at infinity is always present and can not be converted to an angular node.* Thus by beginning with the total number of finite nodes in spherical form for any given quantum number and converting them one at a time to conal and then to planar nodes, all the possible nodal combinations can be realized in a systematic way, as illustrated in Table 2-3. The resulting canonical AO lobes shown in Figure 2-6 occupy the spaces between the nodal surfaces.

In drawing orbital shapes, we should realize that the actual contours of the lobes (i.e., surfaces on which the orbital has a constant value) are defined by mathematical equations. The outer contours shown in Figure 2-6 can be considered to contain about 90% of the electron density but they are really only sketches which depict this outer contour in an approximate fashion. Accurate two-dimensional contour plots of a $(2p)$ and a pair of $(3d)$ AOs generated by computer are shown in Figure 2-7. The three-dimensional contours of the $(2p0) = (2pz)$ and the $(3d0) = (3dz^2)$ orbitals for example can be obtained by spinning the two-dimensional plots around their $z$ axis.

Because the lobes of the $(3d1) = (3dxz)$ are shaped like kidney beans, this orbital is not cylindrically symmetrical about any axis, and one would have to

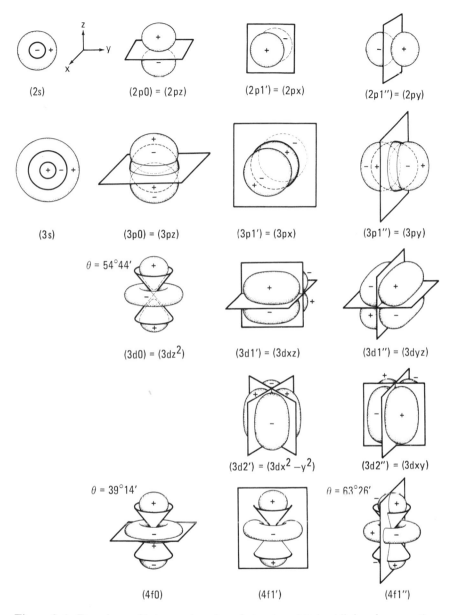

Figure 2-6. Drawings of lobes and nodes of atomic orbitals. All drawings are three-dimensional representations except those for the (2s) and (3s) orbitals. The angles $\theta$ (see Figure 2-1) for the conal nodes are also given. The orbitals are not drawn to scale.

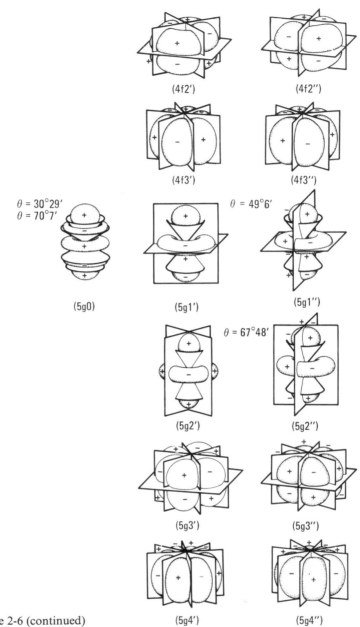

Figure 2-6 (continued)

stack a series of contour plots along *y* which are parallel to the *xz* plane in order to obtain a three-dimensional picture of this orbital. *Instructions can be found in Appendix I for constructing crude three-dimensional models of these orbitals and for a simple device for envisioning molecular orbitals composed of atomic orbitals in the variety of molecular geometries to be discussed in future chapters.*

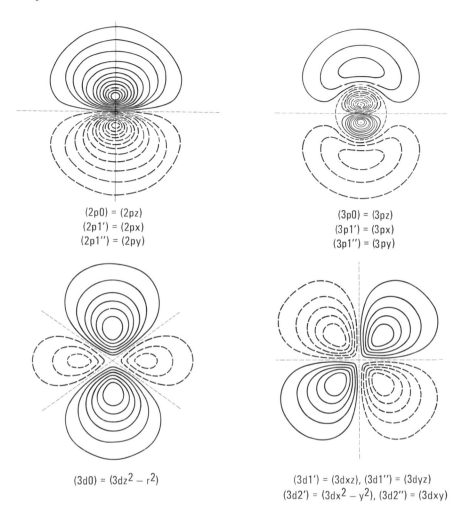

$(2p0) = (2pz)$
$(2p1') = (2px)$
$(2p1'') = (2py)$

$(3p0) = (3pz)$
$(3p1') = (3px)$
$(3p1'') = (3py)$

$(3d0) = (3dz^2 - r^2)$

$(3d1') = (3dxz), (3d1'') = (3dyz)$
$(3d2') = (3dx^2 - y^2), (3d2'') = (3dxy)$

Figure 2-7. Two-dimensional contour plots of some hydrogen orbitals in which each contour has the same electron density. Regions where the wave function is positive are solid contours and where it is negative the contours are dashed lines. The nodes are represented by shorter dashed lines. Note the planar nodes in $(2px)$, $(2py)$, and in the set of four $d$ orbitals; the conal nodes in $(2pz)$, $(3pz)$, $(3dz^2)$, $(3dxz)$, $(3dyz)$; and the spherical node in $(3pz)$, $(3px)$, $(3py)$. Because of the greater density in the smaller $(2p)$ orbital (which is not drawn to scale), the contours are reduced in number by a factor of five compared to the larger $(3p)$ and $(3d)$ orbitals in order to avoid running too many contours into each other.

Table 2-3.  Nodal Surfaces in Canonical Atomic Orbitals

| Principal Quantum Number | Total Number of Nodes[a] | Number of Each Nodal Type | | | Orbital Shape | Orbital Designation |
|---|---|---|---|---|---|---|
| | | Spherical | Conal | Planar | | |
| 1 | 0 | 0 | 0 | 0 | Sphere | $(1s0) = (1s)$ |
| 2 | 1 | 1 | 0 | 0 | Figure 2-6 | $(2s0) = (2s)$ |
| | | 0 | 1 | 0 | Figure 2-6 | $(2p0) = (2pz)$ |
| | | 0 | 0 | 1 | Figure 2-6 | $(2p1') = (2px)$ |
| | | 0 | 0 | 1 | Figure 2-6 | $(2p1'') = (2py)$ |
| 3 | 2 | 2 | 0 | 0 | Figure 2-6 | $(3s0) = (3s)$ |
| | | 1 | 1 | 0 | | $(3p0) = (3pz)$ |
| | | 1 | 0 | 1 | Figure 2-6 | $(3p1') = (3px)$ |
| | | 1 | 0 | 1 | | $(3p1'') = (3py)$ |
| | | 0 | 2 | 0 | | $(3d0) = (3dz^2)$ |
| | | 0 | 1 | 1 | | $(3d1') = (3dxz)$ |
| | | 0 | 1 | 1 | Figure 2-6 | $(3d1'') = (3dyz)$ |
| | | 0 | 0 | 2 | | $(3d2') = (3dx^2 - y^2)$ |
| | | 0 | 0 | 2 | | $(3d2'') = (3dxy)$ |
| 4 | 3 | 3 | 0 | 0 | | $(4s0)$ |
| | | 2 | 1 | 0 | Shaped like $(3p)$'s (Figure 2-6) with one more nodal sphere inside | $(4p0)$ |
| | | 2 | 0 | 1 | | $(4p1')$ |
| | | 2 | 0 | 1 | | $(4p1'')$ |
| | | 1 | 2 | 0 | Shaped like $(3d)$'s (Figure 2-6) with a nodal sphere inside | $(4d0)$ |
| | | 1 | 1 | 1 | | $(4d1')$ |
| | | 1 | 1 | 1 | | $(4d1'')$ |
| | | 1 | 0 | 2 | | $(4d2')$ |
| | | 1 | 0 | 2 | | $(4d2'')$ |
| | | 0 | 3 | 0 | Figure 2-6 | $(4f0)$ |
| | | 0 | 2 | 1 | | $(4f1')$ |
| | | 0 | 2 | 1 | | $(4f1'')$ |
| | | 0 | 1 | 2 | | $(4f2')$ |
| | | 0 | 1 | 2 | | $(4f2'')$ |
| | | 0 | 0 | 3 | | $(4f3')$ |
| | | 0 | 0 | 3 | | $(4f3'')$ |

[(5s), (5p), (5d), and (5f) appear the same as in fourth quantum shell except there is one additional spherical node]

| n | l | | spherical node | | |
|---|---|---|---|---|---|
| 5 | 4 | 4 | 0 | 0 | (5g0) |
| | | 3 | 0 | 1 | (5g1') |
| | | 3 | 0 | 1 | (5g1") |
| | | 2 | 0 | 2 | (5g2') |
| | | 2 | 0 | 2 | (5g2") |
| | | 1 | 0 | 3 | (5g3') |
| | | 1 | 0 | 3 | (5g3") |
| | | 0 | 0 | 4 | (5g4') |
| | | 0 | 0 | 4 | (5g4") |

Figure 2-6

[a] Not including the node at infinity.

## 2.2. Sizes of Canonical AOs

The sizes of orbitals depend on the total number of nodes and the magnitude of the nuclear charge. For the one-electron orbitals of hydrogen, the average orbital radius $\bar{R}$ is

$$\bar{R}_H(nlm) = n^2 \left[ \frac{3}{2} - \frac{l(l+1)}{2n^2} \right] a \qquad (2\text{-}5)$$

where $a$ is the Bohr radius. Here $\bar{R} = \int dV r \phi^2$ the summation of the product of the distance of all points from the origin and the weighting factor $\phi^2 dV$ representing the fraction of the orbital charge in $dV$. The Bohr radius is given by $a = (h/2\pi)^2/me^2 = 5.29177 \times 10^{-9}$ cm and it represents the radius of the circular orbit of lowest energy for the electron in hydrogen as obtained by the classical model. Note that in such formulas $m$ is the mass of the electron and not the quantum number $m$. It is also possible to define a radius $R^*$ within which 90% of the electron is contained:

$$\int_{V(R^*)} \int \int dV[\psi(x, y, z)]^2 = 0.9. \qquad (2\text{-}6)$$

Here $V(R^*)$ denotes the volume inside the sphere with radius $R^*$. In Table 2-4 are collected values of $\bar{R}$, $R^*$, and the ratio $R^*/\bar{R}$ for a range of hydrogen orbitals. Interestingly the ratio decreases with $n$ for a given $l$, and it rises with $l$ for a given $n$ value.

If we increase the nuclear charge $Z$ while retaining only one electron, we

Table 2-4. Various Measures for the Sizes of Orbitals of Hydrogen in Bohr Radii

| Orbital[a] | $R = n^2$ [b] | $\bar{R}$ [c] | $R^*$ [d] | $R^*/\bar{R}$ |
|---|---|---|---|---|
| (1s) | 1 | 1.5 | 2.66 | 1.77 |
| (2s) | 4 | 6 | 9.13 | 1.52 |
| (2p) | 4 | 5 | 7.99 | 1.60 |
| (3s) | 9 | 13.5 | 19.44 | 1.44 |
| (3p) | 9 | 12.5 | 18.39 | 1.47 |
| (3d) | 9 | 10.5 | 15.80 | 1.50 |
| (4s) | 16 | 24 | 33.62 | 1.40 |
| (4p) | 16 | 23 | 32.59 | 1.42 |
| (4d) | 16 | 21 | 30.32 | 1.44 |
| (4f) | 16 | 18 | 25.88 | 1.44 |
| (5s) | 25 | 37.5 | 51.69 | 1.38 |
| (5p) | 25 | 36.5 | 50.67 | 1.39 |
| (5d) | 25 | 34.5 | 48.50 | 1.41 |
| (5f) | 25 | 31.5 | 44.82 | 1.42 |
| (5g) | 25 | 27.5 | 38.52 | 1.40 |

[a] The $m$ quantum number can have any value appropriate to the orbital in question.
[b] Equation 2-9.
[c] Equation 2-5.
[d] Equation 2-7.

proceed to He$^+$, then to Li$^{2+}$, etc. Stronger nuclear attraction is expected to shrink the hydrogenic orbitals. In fact, their $\bar{R}$ and $R^*$ radii are related to those in hydrogen and *both these radii are inversely proportional to the nuclear charge*:

$$\bar{R}_Z(nlm) = \bar{R}_H(nlm)/Z$$

$$R_Z^*(nlm) = R_H^*(nlm)/Z. \tag{2-7}$$

In atoms containing more than one electron, interelectronic repulsions occur. Thus if we consider two electrons having different $n$ values, the electron with the higher $n$ quantum number feels the attractive effect of $Z$ to a lesser extent than the electron with the lower $n$ value not only because it is further from the nucleus but also because the inner electrons repel the outer ones. In an approximate description, the latter effect merely changes the constant $Z$ to a lower value (i.e., to an *effective Z* value) for the outer electron, but it does not change the orbital shapes or their $nlm$ classifications. Thus outer electrons are said to be shielded from the nuclear charge. It is evident from Equation 2-7 that the orbital for such a shielded electron is larger than if the unshielded nuclear charge were acting only on the electron in that orbital.

## 2.3. Energies of Canonical AOs

The electrostatic potential energy $V(r)$ of an electron held to a nucleus of charge $Z$ is given by

$$V(r) = -\frac{Ze^2}{r} \tag{2-8}$$

where $r$ is the distance of the electron from the nucleus. This relationship describes the surface of a three-dimensional potential energy well which is depicted in cross section in Figure 2-8. Since $dV\,\psi^2(x, y, z)$ is proportional to the electronic charge density around the point $x$, $y$, $z$ in the volume $dV$, its contribution to the potential energy is given by $(\psi^2 dV)(-Ze^2/r)$. The potential energy $v(nlm)$ of the entire orbital $\psi = (nlm)$ is therefore

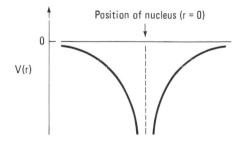

Figure 2-8. Potential energy well for an electron and a positively charged nucleus.

$$v(nlm) = \int\int\int dV [\psi^2(x, y, z)]^2(-Ze^2/r) = -Ze^2/R \qquad (2\text{-}9)$$

where $R$ defined by $R^{-1} = \int\int\int dV r^{-1}\psi^2(x, y, z)$ can be considered as another measure of atom size. This radius, which corresponds to that spherical shell which contains the largest fraction of the electron, also turns out to be the radius of the classical Bohr model for the $(1s)$ orbital.

Since $V(r)$ is negative, $v(nlm)$ is negative, which tells us that the electrostatic force is attractive. The kinetic energy $t(nlm)$ of the electron is always positive and the sum in Equation 2-10 gives us the total orbital energy $\varepsilon(nlm)$:

$$\varepsilon(nlm) = v(nlm) + t(nlm). \qquad (2\text{-}10)$$

If $\varepsilon(nlm)$ is negative, the electron is bound (i.e., it is down in the potential well). If $\varepsilon(nlm) > 0$, the electron is essentially free to move out of the vicinity of the well. Although $t(nlm)$ can be derived rigorously from quantum mechanics, it can also be obtained by considering the Bohr model of an electron as a particle circling the nucleus at a distance $R$. Quantum mechanics requires that the circumference of the path of such a particle in a circular box must contain an integral number of wavelengths, i.e., $\lambda = 2\pi R/n$. A free particle has only kinetic energy, which is given by $t(nlm) = h^2/2m\lambda^2$. Thus we can write Equation 2-11 wherein $\hbar = h/2\pi$:

$$t(nlm) = \left(\frac{\hbar^2}{2mR^2}\right)n^2. \qquad (2\text{-}11)$$

Although we do not do so here, it can be shown that for any stable orbit in a coulombic potential the potential energy $= -2$(kinetic energy). This equality, known as the virial theorem, is also true for quantum mechanical energies. Thus we can say $\frac{1}{2}v(nlm) = -t(nlm)$ and from Equation 2-10 we see that $\varepsilon(nlm) = \frac{1}{2}v(nlm) = -t(nlm)$. Substituting into this equation the expressions for $v(nlm)$ from Equation 2-9 and $t(nlm)$ from Equation 2-11 we have

$$\varepsilon(nlm) = \frac{-Ze^2}{2R} = -\left(\frac{\hbar^2}{2mR^2}\right)n^2 \qquad (2\text{-}12)$$

so that $R = (\hbar^2/me^2)(n^2/Z) = a(n^2/Z)$ where $a$ is the Bohr radius (i.e., $a = R$ when $n$ and $Z = 1$). This means that

$$\varepsilon(nlm) = \frac{-Ze^2}{2a(n^2/Z)} = -\frac{e^2}{2a}\left(\frac{Z}{n}\right)^2 \qquad (2\text{-}13)$$

where $-e^2/2a$ is the total energy of the $(1s)$ orbital electron in hydrogen. Thus the ionization energy $E_H$ or, in other words, the energy required to completely remove this electron from a gaseous hydrogen atom is positive (i.e., $E_H = e^2/2a$) and amounts to 13.6 electron volts (eV). The hydrogen one-electron orbital energies can then be expressed by Equation 2-14:

$$\varepsilon(nlm) = -13.6\left(\frac{Z}{n}\right)^2 \text{ eV}. \qquad (2\text{-}14)$$

This tells us that an increase in nuclear charge leads to tighter binding of the electron while an increase in the $n$ value (i.e., more nodes) leads to looser electron binding. In addition to electron ionization, we note that transitions among orbitals are also possible, and this phenomenon gives rise to the electronic spectrum of hydrogen.

Notice that the energies of hydrogenic orbitals depend only on the quantum number $n$, and not on $l$ or $m$. The energies of orbitals of different $l$ and $m$ values for a given $n$ value are therefore the same in one-electron atoms. Orbitals of the same energy are said to be *degenerate*.

As a consequence of Equation 2-13, the shielded orbital energy is smaller in absolute value (less negative) than an unshielded one. Thus, an electron in a shielded orbital is less tightly bound and therefore has a lower ionization energy. The magnitude of the shielding effect on an orbital depends upon the number of electrons occupying the orbitals which lie between the nucleus and the region of maximal density of the orbital in question. Another important feature of shielding is that it varies with $l$ for a given value of $n$. Consequently there are substantial energy differences among the $s$, $p$, $d$, *etc.*, orbitals within a given quantum shell. It is the shielding effect that gives rise to the orbital occupancy sequence represented by the mnemonic device depicted in Figure 2-9. This ordering of the energy levels of the occupied orbitals turns out to be *nearly* the same in all atoms.

## 2.4.  Hybrid Orbitals

In Chapter 1 we found that in many-electron molecules there are states which we can express by alternative molecular orbital descriptions. We now expand this concept to atomic states. For example, in the Be $1s^2 2s^1 2pz^1$ excited state configuration we can replace the $(2s)$ and $(2pz)$ orbitals by two different

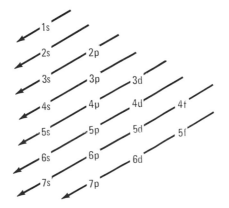

Figure 2-9. Mnemonic device for determining the filling sequence of atomic orbitals.

orbitals of the form

$$(h_1) = A(2s) + B(2pz)$$
$$(h_2) = -B(2s) + A(2pz)$$
(2-15)

where $A$ is arbitrary and $B$ is related to $A$ by $A^2 + B^2 = 1$. Thus we could write $A = \cos\omega$ and $B = \sin\omega$ where $\omega$ is a parameter whose value lies between 0 and $\pi$. It is easily shown that for this excited state of Be, that

$$\rho(x, y, z) = 2(1s)^2 + (2s)^2 + (2pz)^2$$
$$= 2(1s)^2 + (h_1)^2 + (h_2)^2$$
(2-16)

where the coefficient 2 in the first term represents the fact that there are two electrons in the $(1s)$ orbital. The symbol $1s^2$ as a configuration description does not have the same meaning as $(1s)^2$. The latter symbol denotes the square of the $(1s)$ orbital wave function and represents a density. The orbitals $(h_1)$ and $(h_2)$ which are obtained by taking orthogonal linear combinations of different canonical atomic orbitals are called *hybrid orbitals*. Because hybrid atomic orbitals can provide valuable information about molecular orbitals, it is important to discuss some characteristics of hybrid atomic orbitals. It should be realized at the outset, however, that hybrid orbital shapes depend very much on the orbital occupation (i.e., the state) whereas this is not true for canonical orbitals. This will be discussed again later.

Let us first examine what happens when we take linear combinations of a set of three $(p)$ orbitals. To calculate values of the $(2px)$, $(2py)$, and $(2pz)$ orbital amplitudes at various locations $(x, y, z)$ in space, we consider Equations 2-17

$$(2px) = [\pi(2\alpha)^3]^{-1/2}(x/2\alpha)e^{-(r/2\alpha)} = Kxe^{-kr}$$
$$(2py) = [\pi(2\alpha)^3]^{-1/2}(y/2\alpha)e^{-(r/2\alpha)} = Kye^{-kr}$$
$$(2pz) = [\pi(2\alpha)^3]^{-1/2}(z/2\alpha)e^{-(r/2\alpha)} = Kze^{-kr}$$
(2-17)

wherein the constants $K$ and $k$ are the same for $x$, $y$, and $z$ and $\alpha$ has the same meaning as discussed earlier. So that we can conveniently take linear combinations of the $(2p)$ orbitals in Equations 2-17, we will examine some simple properties of vectors. Let $\mathbf{r}$ be the vector which points from the origin to the point $(x, y, z)$ as shown in Figure 2-10a and let $\mathbf{e}$, with components $(\xi, \eta, \zeta)$, be a *unit vector* as depicted in Figure 2-10b. A unit vector $\mathbf{e}$ is one unit long as measured in whatever units we are working with. In Figure 2-10b the unit vector has been placed along the $y$ axis. Then the *projection of* $\mathbf{r}$ *onto* $\mathbf{e}$ denoted by $P_e(\mathbf{r})$ and shown in Figure 2-10c is given by the so-called "dot product" $(\mathbf{e} \cdot \mathbf{r})$ which is defined in Equation 2-18:

$$P_e(\mathbf{r}) = (\mathbf{e} \cdot \mathbf{r}) = \xi x + \eta y + \zeta z.$$
(2-18)

Since the three orbitals in Equations 2-17 are each a function of a single coordinate, Equation 2-18 can be recast into Equations 2-19 which are algebraic expressions for the projections of $r$ onto the three coordinate axes:

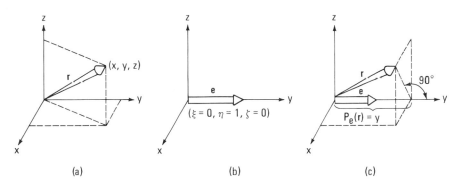

Figure 2-10. Drawings depicting a vector **r** (a), a unit vector **e** (b), and the projection of **r** onto **e** (c).

$$P_{e_x}(\mathbf{r}) = (\mathbf{e}_x \cdot \mathbf{r}) = \xi x, \; P_{e_y}(\mathbf{r}) = (\mathbf{e}_y \cdot \mathbf{r}) = \eta y, \; P_{e_z}(\mathbf{r}) = (\mathbf{e}_z \cdot \mathbf{r}) = \zeta z$$
$$\text{or } P_{e_j}(\mathbf{r}) = (\mathbf{e}_j \cdot \mathbf{r}). \tag{2-19}$$

Here $\mathbf{e}_x = (1, 0, 0)$, $\mathbf{e}_y = (0, 1, 0)$, and $\mathbf{e}_z = (0, 0, 1)$ are the unit vectors in the $x$, $y$, and $z$ directions. Equations 2-19 then permit us to rewrite Equations 2-17 as Equation 2-20:

$$(2pj) = KP_{e_j}(\mathbf{r})e^{-kr} \quad \text{where } j = x, y, z. \tag{2-20}$$

We already know, of course, that $(2px)$, $(2py)$, and $(2pz)$ possess identical shapes and differ only in the directions along which they are oriented. Thus, for example, the $(2px)$ orbital is concentric around $x$ and, by convention, we think of it as pointed along the $+x$ direction when the positive amplitude is in the $+x$ region and the minus amplitude is in the $-x$ region. Similar arguments apply to $(2py)$ and $(2pz)$.

Equation 2-20 actually allows us to think of a $(2p)$ orbital as pointed along any direction we choose, since we can formulate this equation for an arbitrary direction along **e** as in Equation 2-21 where $P_e(\mathbf{r})$ is the projection of $\mathbf{r} = (x, y, z)$ onto $\mathbf{e} = (\xi, \eta, \zeta)$:

$$(2p\mathbf{e}) = KP_e(\mathbf{r})e^{-kr}. \tag{2-21}$$

If we now insert the right-hand side of Equation 2-18 into Equation 2-21 we obtain

$$(2p\mathbf{e}) = K(\xi x + \eta y + \zeta z)e^{-kr}$$
$$= \xi Kxe^{-kr} + \eta Kye^{-kr} + \zeta Kze^{-kr} \tag{2-22}$$

which upon substitution with Equations 2-17 gives

$$(2p\mathbf{e}) = \xi(2px) + \eta(2py) + \zeta(2pz). \tag{2-23}$$

Equation 2-23 says that any linear combination of the orbitals $(2px)$, $(2py)$, and $(2pz)$ (such that $\xi^2 + \eta^2 + \zeta^2 = 1$) gives again an orbital of the $(2p)$ type but pointing in the direction of the unit vector $\mathbf{e} = (\xi, \eta, \zeta)$.

With this result we can rotate *the set of all three p orbitals*, while keeping them orthogonal. Suppose that we have three *mutually orthogonal* unit vectors $\mathbf{e}_1$, $\mathbf{e}_2$, $\mathbf{e}_3$ which are related to $\mathbf{e}_x$, $\mathbf{e}_y$, $\mathbf{e}_z$ by the orthogonal transformation matrix in Equation 2-24:

$$
\begin{array}{c|ccc}
 & \mathbf{e}_x & \mathbf{e}_y & \mathbf{e}_z \\
\hline
\mathbf{e}_1 & T_{11} & T_{12} & T_{13} \\
\mathbf{e}_2 & T_{21} & T_{22} & T_{23} \\
\mathbf{e}_3 & T_{31} & T_{32} & T_{33}
\end{array}
\qquad (2\text{-}24)
$$

wherein the $T_{ik}$ satisfy the relations $T_{i1} T_{k1} + T_{i2} T_{k2} + T_{i3} T_{k3} = 1$ if $i = k$ and $= 0$ if $i \neq k$ (Chapter 1.6). Suppose now that we transform the orbitals $(2px)$, $(2py)$, $(2pz)$ into three $(2p)$ orbitals $(2p\mathbf{e}_1)$, $(2p\mathbf{e}_2)$, $(2p\mathbf{e}_3)$ pointing in the directions $\mathbf{e}_1, \mathbf{e}_2$, and $\mathbf{e}_3$, respectively. We can then express this transformation by the transformation matrix in Equation 2-25:

$$
\begin{array}{c|ccc}
 & (2px) & (2py) & (2pz) \\
\hline
(2p\mathbf{e}_1) & T_{11} & T_{12} & T_{13} \\
(2p\mathbf{e}_2) & T_{21} & T_{22} & T_{23} \\
(2p\mathbf{e}_3) & T_{31} & T_{32} & T_{33}.
\end{array}
\qquad (2\text{-}25)
$$

## 2.5. Hybrid Orbitals through *s-p* Mixing

With this groundwork, we are in a position to consider mixing $(2s)$ and $(2p)$ AOs *in different ratios*. These cases are quite interesting to chemists. Let us mix the $(2s)$ and $(2px)$ orbitals to give a hybrid $(h)$ in Equation 2-26:

$$(h) = A(2s) + B(2px). \qquad (2\text{-}26)$$

Since the canonical orbitals are orthonormal (i.e., $\langle (2s)|(2s) \rangle = 1$, $\langle (2px)|(2px) \rangle = 1$, $\langle (2s)|(2px) \rangle = 0$) we can write Equation 2-27:

$$(h)^2 = A^2 \langle (2s)|(2s) \rangle + B^2 \langle (2px)|(2px) \rangle + 2AB \langle (2s)|(2px) \rangle = A^2 + B^2 \qquad (2\text{-}27)$$

wherein $A^2 + B^2 = 1$ if $(h)$ is to be normalized. If $A^2 = 0.8$ and $B^2 = 0.2$, we say that the hybrid has 80% $(2s)$ and 20% $(2p)$ character. Often $A$ and $B$ are expressed as $A = [a/(a + b)]^{1/2}$, $B = [b/(a + b)]^{1/2}$ where $a$ and $b$ are small integers or some other convenient numbers. The hybrid is then called a $(2s)^a(2p)^b$ hybrid and we can write Equation 2-26 as Equation 2-28:

$$h[(2s)^a(2p)^b] = [a/(a + b)]^{1/2}(2s) + [b/(a + b)]^{1/2}(2p). \qquad (2\text{-}28)$$

A common convention is to abbreviate the orbital character of such a hybrid as $s^a p^b$. *Neither convention is to be confused with orbital occupations.* In Figure 2-11 are contour plots for fourteen $(2s)^a(2px)^b$ carbon hybrids. On the largest

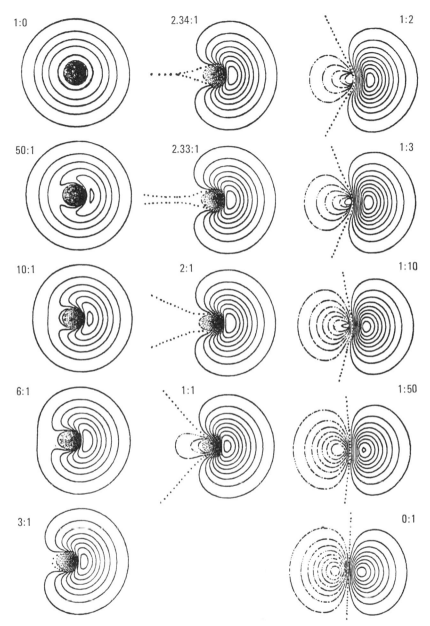

Figure 2-11. Contour plots ($(h) = $ constant) of mixtures of $(s)$ and $(p)$ atomic orbitals showing the ratios $A^2 : B^2$. In these plots solid, dotted, and dashed lines denote regions where $(h) > 0$, $(h) = 0$, and $(h) < 0$, respectively. The absolute value of the smallest value of $(h)$ is 0.05 Bohr$^{-3/2}$ and $\Delta(h)$ has the same value. (Courtesy of Dr. K. Ruedenberg.)

contour, $h$ always has the value of either $+0.05a^{-3/2}$ or $-0.05a^{-3/2}$. The ratio of $a : b = A^2 : B^2$ is given for each contour. Thus this ratio is $1 : 0$ for the pure $(2s)$ and $0 : 1$ for the pure $(2px)$. As more $(2px)$ is mixed into the $(2s)$ (i.e., $50 : 1$, $10 : 1, 6 : 1$, etc.), the orbital becomes increasingly deformed, though it retains the $x$ axis as its axis of symmetry. This deformation or "polarization" with increasing $p$ character is initially accompanied by elongation of the nodal sphere for $(a/b) = 0$. For $(a/b) = 2.34$ the node elongates to a single line in the $-x$ direction. This line for $(a/b) < 2.34$ opens into a cone-like surface, and as $a/b$ decreases still further this surface flattens to a plane which is the nodal plane of the pure $(2px)$ orbital.

For chemically bonded atoms which have more than one AO involved in bonding, it is frequently advantageous to construct hybrid AOs from a canonical AO set. For example, atoms in the second period frequently bind to four equivalent neighboring atoms. In the case of carbon (e.g., in $CH_4$) this set of AOs is comprised of the AOs $(2s)$, $(2px)$, $(2py)$, and $(2pz)$. To help us understand chemical bonding, we distinguish between *core* AOs and *valence* AOs (VAOs). Ionization energies associated with VAOs are low enough to allow involvement of such AOs in chemical bonding while core AOs have higher ionization energies. The canonical carbon VAOs can be linearly combined according to the transformation in Equation 2-29 to form the "hybrid quadruple" or "$(h)$ set":

|          | $(2s)$    | $(2px)$   | $(2py)$   | $(2pz)$    |
|----------|-----------|-----------|-----------|------------|
| $(h_1)$  | $T_{10}$  | $T_{11}$  | $T_{12}$  | $T_{13}$   |
| $(h_2)$  | $T_{20}$  | $T_{21}$  | $T_{22}$  | $T_{23}$   |
| $(h_3)$  | $T_{30}$  | $T_{31}$  | $T_{32}$  | $T_{33}$   |
| $(h_4)$  | $T_{40}$  | $T_{41}$  | $T_{42}$  | $T_{43}.$  |

(2-29)

We know that all the canonical $(nlm)$ orbitals are mutually orthonormal (i.e., $\langle nlm|n'l'm'\rangle = 1$ if $n = n', l = l', m = m'$ and $= 0$ otherwise). For the $(h)$ set to be orthonormal, the transformation in Equation 2-29 must be an orthogonal one, as discussed previously. The $(h)$ set is equivalent to the canonical AO set $(2s)$, $(2px)$, $(2py)$, $(2pz)$ in that any linear combination of the latter set can be expressed as a linear combination of the four members of the $(h)$ set and vice versa.

In what directions do these four orthogonal hybrids "point" in space? As we shall see, this depends on how many of them we "mix" together and in what ratios they are mixed. Let us first examine two $(p)$ orbitals pointing along two arbitrary unit vector directions $\mathbf{e}_1 = (\xi_1, \eta_1, \zeta_1)$ and $\mathbf{e}_2 = (\xi_2, \eta_2, \zeta_2)$ so that

$$(2p\mathbf{e}_1) = \xi_1(2px) + \eta_1(2py) + \zeta_1(2pz)$$
$$(2p\mathbf{e}_2) = \xi_2(2px) + \eta_2(2py) + \zeta_2(2pz).$$

(2-30)

For the overlap integral between these two orbitals, we find

$$\langle(2p\mathbf{e}_1)|(2p\mathbf{e}_2)\rangle = \xi_1\xi_2\langle(2px)|(2px)\rangle + \xi_1\eta_2\langle(2px)|(2py)\rangle$$
$$+ \xi_1\zeta_2\langle(2px)|(2pz)\rangle + \eta_1\xi_2\langle(2py)|(2px)\rangle$$
$$+ \eta_1\eta_2\langle(2py)|(2py)\rangle + \eta_1\zeta_2\langle(2py)|(2pz)\rangle \qquad (2\text{-}31)$$
$$+ \zeta_1\xi_2\langle(2pz)|(2px)\rangle + \zeta_1\eta_2\langle(2pz)|(2py)\rangle$$
$$+ \zeta_1\zeta_2\langle(2pz)|(2pz)\rangle$$

which upon imposition of the condition that these two AOs must be orthogonal simplifies to Equation 2-32:

$$\langle(2p\mathbf{e}_1)|(2p\mathbf{e}_2)\rangle = \xi_1\xi_2 + \eta_1\eta_2 + \zeta_1\zeta_2. \qquad (2\text{-}32)$$

The right-hand side of Equation 2-32 is, of course, the dot product between two directions $\mathbf{e}_1$ and $\mathbf{e}_2$, which we can write as Equation 2-33:

$$\langle(2p\mathbf{e}_1)|(2p\mathbf{e}_2)\rangle = (\mathbf{e}_1 \cdot \mathbf{e}_2) = \cos\gamma_{12} \qquad (2\text{-}33)$$

wherein $\gamma$ is the angle between the two directions. It follows then that if these two $p$ AOs are to be orthogonal (i.e., for Equation 2-33 to be equal to zero), $\gamma$ must be 90°, i.e., $\mathbf{e}_1$ and $\mathbf{e}_2$ must be orthogonal.

Now moving on to our hybrid quadruple, consider any pair [say, $(h_1)$ and $(h_2)$] pointing in the directions $\mathbf{e}_1$ and $\mathbf{e}_2$. We can write normalized expressions for these hybrids as Equation 2-34 (wherein $0 < \alpha < 90°, 0 < \beta < 90°$):

$$(h_1) = \cos\alpha(2s) + \sin\alpha(2p\mathbf{e}_1)$$
$$(h_2) = \cos\beta(2s) + \sin\beta(2p\mathbf{e}_2) \qquad (2\text{-}34)$$

so that all four trigonometric coefficients are positive. Consequently the positive lobe of each hybrid AO $(h_k)$ points in the direction $\mathbf{e}_k$. The parameters $\alpha$ and $\beta$ determine the $(s{-}p)$ mixtures of $(h_1)$ and $(h_2)$, respectively. Proceeding now in a manner analogous to the one above for two orthogonal $(p)$ orbitals we write the overlap integral in Equation 2-35:

$$\langle(h_1)|(h_2)\rangle = \cos\alpha\cos\beta + \sin\alpha\sin\beta(\mathbf{e}_1 \cdot \mathbf{e}_2)$$
$$= \cos\alpha\cos\beta + \sin\alpha\sin\beta\cos\gamma_{12}. \qquad (2\text{-}35)$$

For $(h_1)$ and $(h_2)$ to be orthogonal, Equation 2-35 must be equal to zero, from which Equation 2-36 follows:

$$\cos\gamma_{12} = -\cot\alpha \cdot \cot\beta. \qquad (2\text{-}36)$$

This equation says that when two $(s{-}p)$ hybrids are orthogonal to each other, their degrees of hybridization uniquely determine the angle $\gamma_{12}$ between their directions $\mathbf{e}_1$ and $\mathbf{e}_2$. Moreover, according to Equation 2-36, this angle is always greater than 90° unless at least one of the hybrids has 100% $(p)$ character ($\cos\alpha = 0$ or $\cos\beta = 0$) in which case this $(p)$ orbital forms 90° angles with *all* the other hybrids.

How are the energies of hybrid AOs related to those of their constituent canonical AOs? Since the hybrid AOs are constructed by means of an orthogonal transformation, it can be shown that

$$\varepsilon(h_k) = T_{k0}{}^2\varepsilon(2s) + (T_{k1}{}^2 + T_{k2}{}^2 + T_{k3}{}^2)\varepsilon(2p) \tag{2-37}$$

where $\varepsilon(2p) = \varepsilon(2px) = \varepsilon(2py) = \varepsilon(2pz)$. Since $T_{k0}{}^2 + T_{k1}{}^2 + T_{k2}{}^2 + T_{k3}{}^2 = 1$, we have Equation 2-38:

$$\varepsilon(h_k) = T_{k0}{}^2\varepsilon(2s) + (1 - T_{k0}{}^2)\varepsilon(2p) \tag{2-38}$$

which for an orbital of the type described in Equation 2-26 can be written

$$\varepsilon(h) = A^2\varepsilon(2s) + B^2\varepsilon(2p). \tag{2-39}$$

## 2.6. Equivalent $s^1p^1$ Hybrid Orbitals

We can, of course, have an infinite number of hybrid sets depending on the number of orbitals we linearly combine and on their relative contributions to the hybrid set. For chemical bonding, however, three important cases will now be considered, namely, the formation of *equivalent hybrid sets* of two, three, and four hybrid AOs by linearly combining an ($s$) with one, two, and three ($p$) orbitals, respectively. Equivalent orbitals are defined as being identical in shape but pointing in different directions. For example, a ($2px$) orbital is equivalent to a ($2py$) orbital. Their equivalency is demonstrated by the fact that rotation by 90° about the $z$ axis converts one in to the other. Thus, orbital equivalency is characterized by rotation about a symmetry axis or reflection through a symmetry plane which carries one orbital into another. Equivalent hybrid AOs have identical $s$-$p$ mixing ratios and $\alpha = \beta$ in Equation 2-36. If there are several equivalent hybrids, then $\gamma_{12}$, *the angle between any two of them must be the same.*

Linearly combining an ($s$) with a ($p$) AO gives rise to two equivalent *digonal* hybrid AOs ($d_1$) and ($d_2$). They are called digonal because they lie 180° apart and point in opposite directions. The expressions for our hybrid quadruple now become

$$\begin{aligned}(h_1) &= 2^{-1/2}[(2s) + (2px)] = (d_1)\\[4pt](h_2) &= 2^{-1/2}[(2s) - (2px)] = (d_2)\\[4pt](h_3) &= (2py)\\[4pt](h_4) &= (2pz).\end{aligned} \tag{2-40}$$

The $2^{-1/2}$ coefficient in the equations for ($d_1$) and ($d_2$) normalizes these hybrids (i.e., the sum of the squares of the orbital coefficients must equal 1). That ($d_1$) and ($d_2$) indeed point in opposite directions can be visualized by linearly combining the orbital contours themselves as in Figure 2-12. In this figure it is seen that the positive lobe of the ($2p$) orbital is enhanced by the additive effect

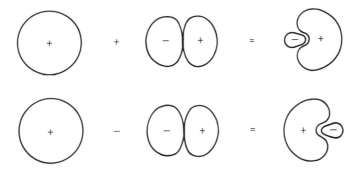

Figure 2-12. Outer contours of digonal orbitals obtained from linearly combining an $(s)$ and a $(p)$ orbital.

of the positive $(2s)$ orbital while the negative lobe of the $(p)$ orbital is partially cancelled by the positive $(2s)$ orbital. To do this accurately, many density contours must be considered rather than just one. It is seen from Figure 2-12 *that the formation of hybrids results in a net localization of electron density in the large lobe.* It can be shown that the expressions in Equations 2-40 require the transformation given in Equation 2-41:

|         | $(2s)$    | $(2px)$    | $(2py)$ | $(2pz)$ |
|---------|-----------|------------|---------|---------|
| $(h_1)$ | $2^{-1/2}$ | $2^{-1/2}$  | 0       | 0       |
| $(h_2)$ | $2^{-1/2}$ | $-2^{-1/2}$ | 0       | 0       |
| $(h_3)$ | 0         | 0          | 1       | 0       |
| $(h_4)$ | 0         | 0          | 0       | 1 .     |

$$(2\text{-}41)$$

The energies of the two identical digonal hybrids are, of course, degenerate and it is easily shown from Equations 2-37 to 2-39 that they lie at the average of the $(2p)$ and $(2s)$ energies (Figure 2-13).

We could, if we wish to do so, reorient our $(h)$ quadruple set along unit

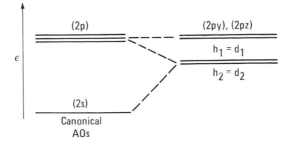

Figure 2-13. Diagram showing the relationship of the energies of the canonical and hybrid AOs when an $(s)$ AO is mixed with a $(p)$ AO in a $1:1$ ratio.

vectors $\mathbf{d}_1$, $\mathbf{d}_2$, $\mathbf{d}_3$, $\mathbf{d}_4$, which are generated pictorially in Figure 2-14 by linearly combining $(px)$, $(py)$, and $(pz)$ AOs. From Figure 2-14a and b we conclude that $(d+)$ and $(d-)$ point along $\mathbf{d}_1$ and $\mathbf{d}_2$, respectively. In order for the pure $(p)$ hybrids $(2p\mathbf{d}_3)$ and $(2p\mathbf{d}_4)$ to be at right angles to the new digonal hybrids $(d_+)$ and $(d_-)$ (see previous discussion), they must be formed as shown in Figure 2-14c and d. The unit vectors $\mathbf{d}_1$, $\mathbf{d}_2$, $\mathbf{d}_3$, $\mathbf{d}_4$ are seen from Figure 2-14a–d to have the directions $\mathbf{d}_1 = (-2^{-1/2}, 2^{-1/2}, 0)$, $\mathbf{d}_2 = (2^{-1/2}, -2^{-1/2}, 0) = -\mathbf{d}_1$, $\mathbf{d}_3 = (1/2, 1/2, 2^{-1/2})$, $\mathbf{d}_4 = (-1/2, -1/2, 2^{-1/2})$. These directions are the coefficients of the $(2p)$ orbitals which allow our reoriented $p$ hybrids $(2p\mathbf{d}_1)$, $(2p\mathbf{d}_2)$, $(2p\mathbf{d}_3)$, $(2p\mathbf{d}_4)$ in Figure 2-14 to be normalized. Thus the sum of the squares of the coefficients is 1 in all cases. That these coefficients are consistent with the geometrical constraints of our reorientation can be seen by considering Figure 2-14a as an example. Here we are taking the linear combination of two vectors, each of which is one unit in length from the origin to either end. Taking the positive lobe as our reference, this unit vector has the coordinates $(-2^{-1/2}, 2^{-1/2}, 0)$ as shown in Figure 2-15. In Figure 2-16 is depicted a composite of the relative orientations of the unit vectors $\mathbf{d}_1$, $\mathbf{d}_2$, $\mathbf{d}_3$, $\mathbf{d}_4$ along which lie the reoriented $(s^1 p^1)$ hybrids $(h_1)$, $(h_2)$, $(h_3)$, $(h_4)$ shown in Equation 2-42:

$$(h_1) = (d_+) = 2^{-1/2}[(2s) + (2p\mathbf{d}_1)]$$

$$(h_2) = (d_-) = 2^{-1/2}[(2s) + (2p\mathbf{d}_2)]$$

$$(h_3) = (2p\mathbf{d}_3)$$

$$(h_4) = (2p\mathbf{d}_4).$$

$$(2\text{-}42)$$

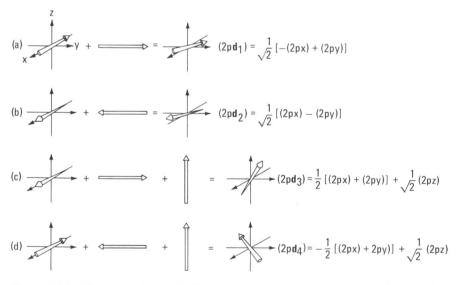

Figure 2-14. The orientation of the full set of $2p$ hybrids along the new unit vectors $\mathbf{d}_1$, $\mathbf{d}_2$, $\mathbf{d}_3$, $\mathbf{d}_4$.

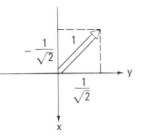

Figure 2-15. The orientation of the unit vector $\mathbf{d}_1$ in the $xy$ plane.

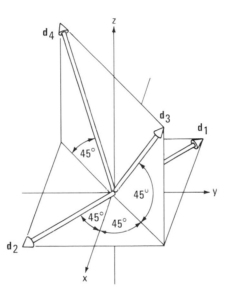

Figure 2-16. The orientation of the unit vectors $\mathbf{d}_1, \mathbf{d}_2, \mathbf{d}_3, \mathbf{d}_4$ in space.

Here the positive lobes of the $(s^1p^1)$ digonal hybrids point away from each other in the $\mathbf{d}_1$ and $\mathbf{d}_2$ directions, and the two pure $(p)$ orbitals lie along $\mathbf{d}_3$ and $\mathbf{d}_4$.

## 2.7. Equivalent $sp^2$ Hybrid Orbitals

If we mix $(2s)$ with *two* $(p)$ orbitals, say, $(2px)$ and $(2py)$, three hybrid AOs are produced whose axes lie in the $xy$ plane. That the axes of such a hybrid set lie in the $xy$ plane is reasonable since the $(2px)$ and $(2py)$ contours when algebraically combined with the $(2s)$ must give hybrid AOs which are symmetrical with respect to this plane which cuts through them. If the linear combinations are carried out so that the resulting three hybrids are identical [i.e., they each contain $1/3$ $(2s)$ character and $2/3$ $(2p)$ character], the fact that they are identical requires that such $(s^1 p^2)$ orbitals point to the vertices of an equilateral triangle. Hence they are called *trigonal* hybrids. To demonstrate pictorially the linear combinations which give rise to hybrids more complicated than digonal ones is difficult and this will not be done here. The orthogonal matrix which transforms the canonical AOs into the *more localized* trigonal hybrids is given in Equation 2-43:

| | (2s) | (2px) | (2py) | (2pz) | Explicit Expressions |
|---|---|---|---|---|---|
| $(h_1)$ | $\dfrac{1}{\sqrt{3}}$ | $\sqrt{\dfrac{2}{3}}$ | 0 | 0 | $\dfrac{1}{\sqrt{3}}(2s) + \sqrt{\dfrac{2}{3}}(2px) = (tr_1)$ |
| $(h_2)$ | $\dfrac{1}{\sqrt{3}}$ | $\dfrac{-1}{\sqrt{6}}$ | $\dfrac{1}{\sqrt{2}}$ | 0 | $\dfrac{1}{\sqrt{3}}(2s) - \dfrac{1}{\sqrt{6}}(2px) + \dfrac{1}{\sqrt{2}}(2py) = (tr_2)$ |
| $(h_3)$ | $\dfrac{1}{\sqrt{3}}$ | $\dfrac{-1}{\sqrt{6}}$ | $\dfrac{-1}{\sqrt{2}}$ | 0 | $\dfrac{1}{\sqrt{3}}(2s) - \dfrac{1}{\sqrt{6}}(2px) - \dfrac{1}{\sqrt{2}}(2py) = (tr_3)$ |
| $(h_4)$ | 0 | 0 | 0 | 1 | $(2pz)$. |

$$(2\text{-}43)$$

These hybrid AOs can also be written in more general form as in Equation 2-44:

$$(h_1) = \frac{1}{\sqrt{3}}(2s) + \sqrt{\frac{2}{3}}(2p\mathbf{f}_1) = (tr_1)$$

$$(h_2) = \frac{1}{\sqrt{3}}(2s) + \sqrt{\frac{2}{3}}(2p\mathbf{f}_2) = (tr_2)$$

$$(h_3) = \frac{1}{\sqrt{3}}(2s) + \sqrt{\frac{2}{3}}(2p\mathbf{f}_3) = (tr_3)$$

$$(h_4) = (2pz)$$

$$(2\text{-}44)$$

wherein the $(2p\mathbf{f}_k)$ orbitals are defined in the case at hand by Equation 2-45:

$$(2p\mathbf{f}_1) = (2px)$$

$$(2p\mathbf{f}_2) = -\frac{1}{2}(2px) + \frac{1}{2}\sqrt{3}(2py)$$

$$(2p\mathbf{f}_3) = -\frac{1}{2}(2px) - \frac{1}{2}\sqrt{3}(2py).$$

$$(2\text{-}45)$$

The $(2p\mathbf{f}_k)$ orbitals point in the directions $\mathbf{f}_1 = (1, 0, 0)$, $\mathbf{f}_2 = (-1/2, 3^{1/2}/2, 0)$, $\mathbf{f}_3 = (-1/2, -3^{1/2}/2, 0)$. These vectors are of unit length so that the $(2p\mathbf{f}_k)$ are normalized. As shown in Figure 2-17a, the unit vector directions are at 120° to one another as are the $(sp^2)$ hybrids which point along these directions. The energy of each of the trigonal hybrids relative to the canonical AOs is depicted in Figure 2-17b.

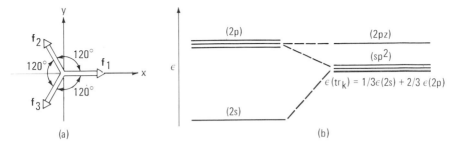

Figure 2-17. Orientation (a) and energies (b) of the $(sp^2)$ trigonal hybrid set of orbitals.

## 2.8. Equivalent $sp^3$ Hybrid Orbitals

Mixing an $(s)$ orbital equally with all three $(p)$ orbitals gives rise to four identical hybrid AOs. It might be thought that four such hybrids could be pointed to the corners of either a square or a tetrahedron. The square can be ruled out, however, from the fact that $\gamma_{12}$ (Equation 2-36) between any two hybrids is not the same. The square can also be eliminated on simple geometrical considerations. To see this let us take the hybridization process in a stepwise manner. Recall that mixing an $(s)$ with one $(p)$ orbital, namely $(px)$, gives us digonal hybrids pointed in opposite directions along $x$. Mixing in an additional $(p)$ orbital, namely $(py)$, with the digonal AOs leads to trigonal hybrids whose axes for geometrical reasons must lie in the $xy$ plane. If we now mix into the $(sp^2)$ set the $(pz)$ orbital, whose axis lies *perpendicular* to the axes of the trigonal AOs, it is impossible to obtain four hybrids whose axes are all in the $xy$ plane since the coefficient of $(pz)$ would be necessarily zero. The only other choice which preserves the *identical* character of all four hybrids is to require the hybrids to point to the vertices of a tetrahedron. The orthogonal transformation of the canonical orbitals which gives such tetrahedral hybrid $(th)$ AOs is given in Equation 2-46:

|         | $(2s)$        | $(2px)$         | $(2py)$         | $(2pz)$         | Explicit Expressions                                                                       |
|---------|---------------|-----------------|-----------------|-----------------|--------------------------------------------------------------------------------------------|
| $(h_1)$ | $\frac{1}{2}$ | $\frac{1}{2}$   | $\frac{1}{2}$   | $\frac{1}{2}$   | $\frac{1}{2}(2s) + \frac{1}{2}(2px) + \frac{1}{2}(2py) + \frac{1}{2}(2pz) = (th_1)$       |
| $(h_2)$ | $\frac{1}{2}$ | $-\frac{1}{2}$  | $-\frac{1}{2}$  | $\frac{1}{2}$   | $\frac{1}{2}(2s) - \frac{1}{2}(2px) - \frac{1}{2}(2py) + \frac{1}{2}(2pz) = (th_2)$       |
| $(h_3)$ | $\frac{1}{2}$ | $\frac{1}{2}$   | $-\frac{1}{2}$  | $-\frac{1}{2}$  | $\frac{1}{2}(2s) + \frac{1}{2}(2px) - \frac{1}{2}(2py) - \frac{1}{2}(2pz) = (th_3)$       |
| $(h_4)$ | $\frac{1}{2}$ | $-\frac{1}{2}$  | $\frac{1}{2}$   | $-\frac{1}{2}$  | $\frac{1}{2}(2s) - \frac{1}{2}(2px) + \frac{1}{2}(2py) - \frac{1}{2}(2pz) = (th_4)$.      |

$$(2\text{-}46)$$

These hybrids can also be expressed as Equation 2-47:

$$(th_k) = \tfrac{1}{2}(2s) + \tfrac{1}{2}\sqrt{3}(2p\mathbf{g}_k) \qquad (2\text{-}47)$$

using the four $(2p)$ orbitals in Equation 2-48 oriented along the four unit vectors $\mathbf{g}_k = (\xi_k, \eta_k, \zeta_k)$

$$(2p\mathbf{g}_k) = \xi_k(2px) + \eta_k(2py) + \zeta_k(2pz), \qquad (2\text{-}48)$$

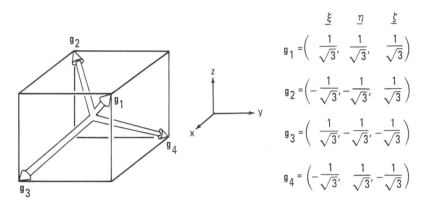

Figure 2-18. Orientation and coordinates of the four unit vectors along which lie the $(sp^3)$ hybrid AOs $(th_k)$.

which point to the corners of a regular tetrahedron as given by their coordinates in Figure 2-18.

Shortly we shall consider the relationships among the various hybrid quadruple sets. To visualize the relationships between the tetrahedral and trigonal hybrids, we shall first briefly discuss an alternative set of hybrids $(\hat{th}_k)$ using the definitions in Equation 2-49:

$$(\hat{th}_k) = \tfrac{1}{2}(2s) + \tfrac{1}{2}\sqrt{3}(2p\hat{g}_k)$$

$$(2pg\hat{g}_k) = \xi(2px) + \eta(2py) + \zeta(2pz) \tag{2-49}$$

wherein the new directional unit vectors $\hat{g}_k = (\xi_k, \eta_k, \zeta_k)$ are given by Equation 2-50:

$$\hat{g}_1 = \tfrac{1}{3}\sqrt{8}\mathbf{f}_1 + \tfrac{1}{3}\mathbf{e}_z$$

$$\hat{g}_2 = \tfrac{1}{3}\sqrt{8}\mathbf{f}_2 + \tfrac{1}{3}\mathbf{e}_z$$

$$\hat{g}_3 = \tfrac{1}{3}\sqrt{8}\mathbf{f}_3 + \tfrac{1}{3}\mathbf{e}_z \tag{2-50}$$

$$\hat{g}_4 = -\mathbf{e}_z .$$

Here the new vectors $\hat{g}_k$ are expressed in terms of the three previously discussed trigonal unit vectors $\mathbf{f}_k$ in the $xy$ plane, and the $z$ axis unit vector $\mathbf{e}_z$. Thus the $\hat{g}_k$ point in the directions shown in Figure 2-19 as is seen from the sole contribution of $-\mathbf{e}_z$ to $\hat{g}_4$ and the positive contributions of $\mathbf{e}_z$ to the trigonal unit vectors $\mathbf{f}_k$. By inserting into Equation 2-50 the values of $\mathbf{f}_k$ given earlier we obtain the coordinates shown in Figure 2-19 for the $\hat{g}_k$. The angle between any pair of $\hat{g}_k$ is the tetrahedral angle of $109°28'$ as can be demonstrated by taking the dot product between any two unit vectors:

$$\cos \gamma_{jk} = (\hat{g}_j \cdot \hat{g}_k) = -\tfrac{1}{3} \text{ for } jk = 12, 13, 14, 23, 24, 34. \tag{2-51}$$

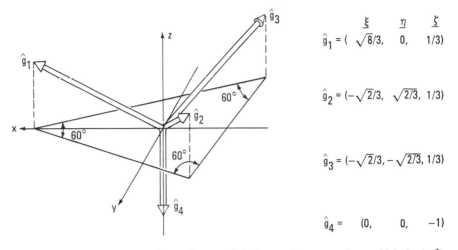

$$\hat{g}_1 = (\begin{array}{ccc} \xi & \eta & \zeta \\ \sqrt{8/3}, & 0, & 1/3 \end{array})$$

$$\hat{g}_2 = (-\sqrt{2/3}, \ \sqrt{2/3}, \ 1/3)$$

$$\hat{g}_3 = (-\sqrt{2/3}, -\sqrt{2/3}, 1/3)$$

$$\hat{g}_4 = (0, \ \ \ \ 0, \ \ -1)$$

Figure 2-19. Orientation and coordinates of the four unit vectors along which the ($sp^3$) hybrid AOs ($\hat{ih}_k$) lie.

For example, $(\hat{\mathbf{g}}_1 \cdot \hat{\mathbf{g}}_2) = -4/9 + 0 + 1/9 = -3/9 = -1/3$. Furthermore, it is clear from Equation 2-47 that these $sp^3$ hybrids have 25% ($s$) and 75% ($p$) character. The only difference between the ($\hat{ih}_k$) and the ($th_k$) hybrids examined earlier is that the two hybrid sets are oriented differently with respect to the Cartesian axes.

## 2.9. Intermediate Hybrid Orbitals

It is often necessary for bonding purposes to consider hybrids which are *intermediate* in character between the *equivalent* hybrids just discussed. Intermediate hybrids can be visualized as continuous transitions from one orbital set to another, e.g., ($s$) plus added ($p$) character until ($sp$) is reached, ($sp$) plus ($p$) character to ($sp^2$), and ($sp^2$) plus ($p$) to ($sp^3$). Such transitions can be accomplished without violating orthogonality conditions. As an example, we consider the transition from ($2s$) plus ($2px$) to ($sp$). We can generalize the transformation of ($2s$) and ($2px$) to the digonal hybrids ($h_+$) and ($h_-$) given earlier in Equations 2-40 and 2-41 by substituting $A$ and $B$ as shown in Equation 2-52:

|                       | ($2s$) | ($2px$) | ($2py$) | ($2pz$) |
|-----------------------|--------|---------|---------|---------|
| ($h_1$) = ($h_+$)     | $A$    | $B$     | 0       | 0       |
| ($h_2$) = ($h_-$)     | $B$    | $-A$    | 0       | 0       |
| ($h_3$)               | 0      | 0       | 1       | 0       |
| ($h_4$)               | 0      | 0       | 0       | 1       |

$$(2\text{-}52)$$

As long as $A^2 + B^2 = 1$, this transformation is orthogonal for any values of $A$ and $B$. Assuming $A > 0$ and $B > 0$, $(h_+)$ points in the $+x$ direction and $(h_-)$ points in the $-x$ direction, and the two hybrids always remain 180° apart. Furthermore, the $(s)$ and $(p)$ characters of $(h_+)$ are the converse of those of $(h_-)$. Thus if $(h_+)$ has 80% $(s)$ character and 20% $(p)$ character, then $(h_-)$ possesses 20% $(s)$ character and 80% $(p)$ character, and the shapes of these orbitals as well as their energies reflect these differences. The transformation in Equation 2-52 affords the possibility of a continuous transition from $(2s)$ plus $(2px)$ to the digonal $(d_+)$ and $(d_-)$ $(s^1p^1)$ set discussed earlier. For $A = 0$, $B = 1$ we have $(h_-) = (2s)$ and $(h_+) = (2px)$. As $A$ increases to positive values (and $B$ decreases concomitantly to preserve orthogonality) $(h^-)$ gains $(p)$ character and $(h_+)$ accumulates $(s)$ admixture until for $A = B = 2^{-1/2}$ both hybrids assume that same shape and energy and they become $(d_+)$ and $(d_-)$.

In a similar manner, transitions between other regular hybrid sets can occur. Examples involving an $(s)$ and three $(p)$ orbitals are pictorially represented in Figure 2-20. In (a) and (h) of this figure are shown two orientations of the digonal set and in (e) and (f) are depicted two orientations of the tetrahedral set. The transition from digonal (a) to trigonal (c) involves bending the two digonal orbitals $(d_+)$ and $(d_-)$ in (a) backwards in the $xy$ plane to become $(h_2)$ and $(h_3)$ in the intermediate (b). Meanwhile, $(p_1)$ gains $(s)$ character [at the expense of $(d_-)$ and $(d_+)$] to become $(h_1)$ in (b). When the angle $\alpha$ in (b) reaches 120°, the trigonal hybrids $(tr_k)$ in (c) are born. If we now imagine that the $(tr_k)$ in (c) are bent upwards in the $+z$ direction to give us $(h_1)$, $(h_2)$, $(h_3)$ in (d), $(p_2)$ in (c) reverses its direction and gains $(s)$ character to become $(h_4)$ in (d). When $\alpha + 90°$ in (d) opens up to 109°28′ we have the $(th_k)$ in (e). One can also imagine a transition from the digonal set in (h) to the tetrahedral set in (f) via the intermediate in (g). Here $(d_-)$ and $(d_+)$ (which are 180° apart) are bent down [at a faster rate than the opening of the angle between $(p_1)$ and $(p_2)$]. The orbitals $(p_1)$ and $(p_2)$ gain $(s)$ character [at the expense of the $(d_+)$ and $(d_-)$ hybrids] until the angles $\alpha$ and $\beta$ in (g) become equal (109°28′) in (f).

# Summary

We have learned how to obtain the nodal and lobal configurations (i.e., the shapes) of canonical AOs by converting spherical to conal and planar nodes. We introduced three different measures of orbital sizes (i.e., $R$, $\bar{R}$ and $R^*$) and briefly examined factors that affect the energies of canonical AOs. We found that the device in Figure 2-9 gives us a simple way for remembering the sequence in which AOs are occupied by electrons for the vast majority of atoms and ions. We then saw how hybrid AO sets can be formed from canonical AOs. When such hybrid AOs are formed by equally mixing, say, a pair of $(2p)$ AOs, two equivalent $(2p)$ AOs are produced which may be oriented in space along any set of mutually orthogonal unit vector directions.

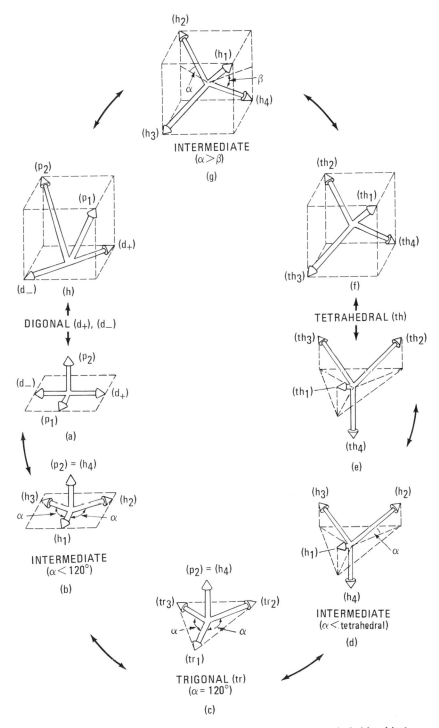

Figure 2-20. Pictorial representation of progressions from one hybrid orbital set to another.

By similarly mixing an $(s)$ AO with one or more $(p)$ AOs, hybrid sets of the digonal, trigonal, and tetrahedral types are formed which, in addition to possessing orientational flexibility, also localize electron density to a greater extent than their canonical parent AOs. Finally, we addressed the possibility of unequal mixing of $(s)$ and $(p)$ AOs and discovered that equivalent hybrid sets are related by series of intermediate hybrids.

PROBLEMS

1. Sketch all the $6h$ atomic orbitals and label the $\theta$, $\phi$, and $r$ nodes.

2. How many spherical nodes (not counting the node at infinity) are present in the $(5d0)$, $(5f1'')$, and $(6p1')$ orbitals?

3. Make a plot of $V(r)$ in eV as a function of $r$ for H and $Li^{2+}$. Label the $r$ axis with the appropriate units of length.

4. Calculate the potential energy and the total energy for the $1s^02s^1$ and the $1s^02pz^1$ states of the hydrogen atom.

5. Verify Equation 2-36.

6. Justify Figure 2-13 from Equations 2-37 to 2-39. In a similar manner justify Figure 2-17b.

7. Show that Equations 2-40, 2-42, and 2-44 are orthogonal for any pair of functions in each.

8. Verify Equation 2-46 by substituting Equation 2-47 and making use of the values of $\mathbf{g}_k$ in Figure 2-18.

9. Show from trigonometric considerations that the $\mathbf{f}_k$ $(k = 1-3)$ for the $(sp^2)$ hybrids stem from the geometric relationship of these orbitals.

10. Sketch the $(1s)$ orbital of hydrogen in HF and the $(2p)$ AOs of the fluorine with one of the $(2p)$ AOs pointing toward the hydrogen. Indicate which $(2p)$ orbital(s) is orthogonal to the hydrogen $(1s)$.

11. What are $\alpha$ and $\beta$ in Equation 2-34 for digonal hybrids?

12. Verify the second line of Equation 2-16.

# CHAPTER 3

# Diatomic Molecules

Because homonuclear diatomic molecules are the simplest of all molecules, we can use them to illustrate some important concepts which are common to all molecules, namely, molecule formation and molecular bonding. In this chapter we introduce the generator orbital (**GO**), which is used here and in subsequent chapters as a device to generate in a pictorial way delocalized and localized molecular orbital (MO) pictures for a variety of molecules. The **GO** is also employed to generate the normal vibrational modes of these molecules.

## 3.1. Molecule Formation and Motions

If the centers of atoms A and B in a diatomic molecule are separated by a finite distance $R$, they experience internuclear repulsion, attraction between the nucleus of one and the electrons of the other, and interelectronic repulsion. In addition to these electrical forces there is a "kinetic force" due to the $R$-dependence of the kinetic energy. All these forces combine to create an effective potential $U(R)$. A repulsive potential [for which $U(R) > 0$ everywhere] is depicted in Figure 3-1a. In this case two atoms held at $R_1$ and then released will move apart. At the same time the potential energy $U(R_1)$ which the system had at $R_1$ will be converted to kinetic energy until at $R(\infty)$ all of the potential energy is converted to kinetic energy and the atoms separate with respect to one another with a kinetic energy $(mv^2/2) = U(R_1)$. For all intents and purposes $U(\infty)$ corresponds to $R \geq$ to about $10^{-4}$ cm. If, on the other hand, two atoms separated by $R(\infty)$ approach one another with a kinetic energy corresponding to $U(R_1)$, they will decrease their relative velocity until at $R_1$ they are halted. However, they immediately fly apart again since the potential curve describing their interaction is overall repulsive. The curve in Figure

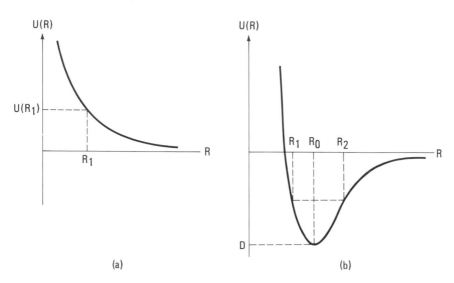

Figure 3-1. Potential energy $[U(R)]$ as a function of distance $(R)$ for a repulsive (a) and an attractive (b) interaction between two atoms. Since the direction of approach of the atoms is arbitrary, the two-dimensional potential diagrams should be spun around the $U(R)$ axis to visualize the potential energy surface.

3-1b describes a potential function which for a range of internuclear distances is attractive $[U(R) < 0]$. That is, a pair of atoms held at an internuclear distance of $R_0$ is unable to move apart since the atoms have no kinetic energy to expend in climbing the wall in either direction. In such a situation we say that $R_0$ is the *equilibrium distance* or the *equilibrium bond length* of the molecule. For a chemical bond of average strength, $R_0$ is about 0.8 to 0.9 times the sum of the atomic radii as defined by $R(nlm) = an^2/Z^*_{nlm}$ where $Z^*_{nlm}$ is the effective nuclear charge as a result of shielding effects. Weak bonds are generally longer and strong bonds shorter.

Suppose now that we impart motion to the two atoms at $R_0$ such that they move toward one another in Figure 3-1b with a kinetic energy of $mv_0^2/2$. This kinetic energy is transformed into potential energy until the atoms stop at $R_1$ where $U(R_1) - U(R_0) = mv_0^2/2$. They then accelerate up to $R_0$ and then decelerate up to $R_2$ whereupon they go back to $R_0$ and the whole cycle repeats itself over and over again. This oscillation constitutes a molecular vibration. Of course, if the kinetic energy initially imparted exceeds the energy difference $U(\infty) - U(R_0) = -U(R_0) = D$, then the atoms will fly apart to infinity. The energy $D$ is called the dissociation energy. The stronger the bond, the larger its dissociation energy. Bonds which are weak, average, and strong have dissociation energies of about 2, 5, and 8 eV, respectively. Weak bonds tend to be easily stretched and compressed compared to strong bonds and their oscillations also tend to be larger than those of strong bonds. These characteristics are reflected in Figure 3-2 in which the potential well is seen to be wider than in Figure 3-1b. Molecular vibrations tend to be quite small

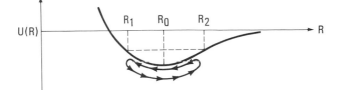

Figure 3-2. A potential well for a relatively weakly bonded diatomic molecule. The amplitude of the vibration is seen to be comparatively large.

(ca. 0.1 Å, 1 Å $= 10^{-8}$ cm) relative to bond lengths, which are on the order of one or two angstroms. It should be realized that molecular vibrations are independent of any translational or rotational motions of the molecules as a whole.

What is the origin of the potential well for stable molecules? In the classical particle picture, electrons whirl around the two fixed nuclei just as they do around an atom at rest. For the purposes of this discussion let us make the simplifying assumption that the two nuclei are at rest separated by $R_0$. The electronic motion then determines $U(R_0)$ which is equal to the total (i.e., kinetic plus potential) energy of all electronic motions plus the internuclear repulsion. This is summarized in Equation 3-1:

$$U(R_0) = E_{el}(R_0) + \frac{e^2 Z_A Z_B}{R_0} \tag{3-1}$$

wherein $eZ_A$ and $eZ_B$ are the nuclear charges. This relationship is also true for $R \neq R_0$ if the nuclei are held at $R$, except that $R_0$ in Equation 3-1 is replaced by $R$. The total electronic energy $E_{el}(R_0)$ must be the dominant contribution to $D = -U(R_0)$ at $R_0$, since the repulsion term $e^2 Z_A Z_B / R_0$ is always positive, whereas $E_{el}(R_0)$, of course, must be negative for stable molecules.

## 3.2. Generator and Molecular Orbitals

The electronic structure and energy of a molecule is determined by the occupied MOs. What do MOs look like? Like AOs, they possess lobes and nodes. The shapes of the nodes of AOs are a consequence of the *spherical* symmetry of the potential energy of an electron around one nucleus. Because molecules have more than one nucleus, the potential energy of an electron will not have spherical symmetry and hence the patterns of its MO nodes and lobes will be less than spherically symmetrical. For additional discussion of precisely what is meant by symmetry and its connection to generator and molecular orbitals, see Appendix I.

In order to obtain the nodal patterns of MOs, we employ a device which consists of *placing an imaginary AO at the center of a molecule. For a diatomic molecule this point is midway between the nuclei. We then consider its nodal*

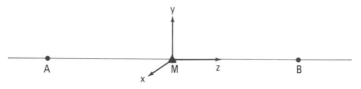

Figure 3-3. Axis system for **GO**s at M, the center of symmetry of a homonuclear diatomic potential provided by nuclei A and B.


*pattern*. By "imaginary" we mean that this AO exists only in our minds. We will refer to these imaginary AOs as *generator orbitals* or **GO**s. Before proceeding to the generation of diatomic MO nodal patterns, let us examine the behavior of some **GO**s in the diatomic potential. Imagine a $(2px)$, a $(2py)$, and a $(2pz)$ AO at the center M in Figure 3-3 of a homonuclear diatomic molecule. From now on, we will generally assume that the coordinates of the **GO**s are oriented so that the $z$ axis lies along the internuclear axis. We will also frequently use a triangle to denote the **GO** center when that location is not obvious. It is clear that both the $(2px)$ and $(2py)$ **GO**s feel the same environment from the two nuclei but that the $(2pz)$ **GO** must feel this environment differently since its lobes are directed toward the nuclei rather than being oriented in a plane perpendicular to the internuclear axis. Thus the energy degeneracy of these three orbitals in the atom is only partially retained in the diatomic molecule. Similarly the $(3d0)$, $(3d1')$, $(3d1'')$, $(3d2')$, and $(3d2'')$ **GO**s lie at three different energies in the diatomic potential, whereas they are fivefold degenerate in the atom: the degenerate pairs are now $(3d1')$, $(3d1'')$ and $(3d2')$, $(3d2'')$. The distinction between different values of $m$ is thus an important one energetically and we employ these symmetry labels for both **GO**s and MOs. In diatomics, orbitals with $m = 0, 1, 2$ are said to be of the $\sigma$, $\pi$, and $\delta$ type, respectively. We will also use these Greek symbols to designate the symmetries of orbitals in polyatomics.

Let us apply the **GO** concept to the generation of the MOs in $N_2$. Each of the A and B atoms of $N_2$ contains a set of $(1s)$, $(2s)$, $(2px)$, $(2py)$, and $(2pz)$ AOs. A fundamental rule is that *we must construct as many MOs as there are AOs available for use*. Recall that a similar rule applies to hybrids in that the number of hybrid AOs must equal the number of canonical AOs we linearly combine. Since we have a total of ten AOs in $N_2$, we can form ten MOs. Using the coordinate system in Figure 3-4, we first place the simplest possible **GO**, namely a $(1s)$ AO, at the molecular center. The advantages of directing the $z$ axes along the internuclear bond axis in the manner shown will become clear later. If we now place a $(1s)$ AO on each nitrogen, we find that the $(1s)$ **GO** has the same symmetry *in the diatomic potential* as a pair of identical $(1s)$ orbitals, one on each nitrogen. To understand that this is so, imagine a point $q$ anywhere in the $(1s)$ **GO**. As shown in Figure 3-5, in which the orbitals are shown in cross-sectional views, an equivalent point $q'$ for that point $q$ can be found by rotating $q$ around the $z$ axis by any angle, by reflecting $q$ through the $xy$ plane, by reflecting $q$ through any plane containing the

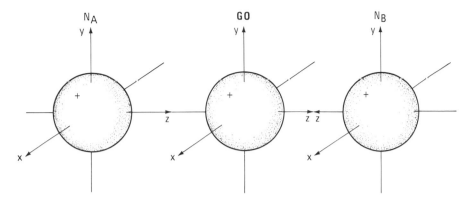

Figure 3-4. A sketch of a (1s) **GO** and two (1s) nitrogen AOs in $N_2$.

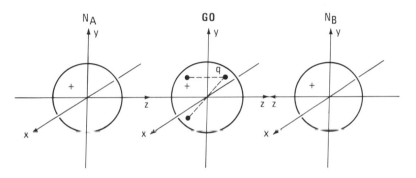

Figure 3-5. A depiction of the identity of the symmetries of a (1s) **GO** and a pair of (1s) AOs of the diatomic molecule $N_2$ in the field of the two nuclei.

molecular axis, or by inverting it through the molecular center. At these equivalent pairs of points, the (1s) **GO** has the same value. By inspection it can be seen that the same is true for the combination of the nitrogen (1s) orbitals in which each nitrogen AO has the same size and sign. In Figure 3-6 we see that the (1s) **GO** also has the same symmetry in the diatomic potential as a pair of identical (2s) orbitals placed on the nitrogens. When AOs possess spherical nodes, we will always place the same sign in the outer lobe as we have in the **GO**. We also see that the symmetry of the (1s) **GO** is the same as that of the pair of nitrogen (2pz) AOs, as is shown in Figure 3-7a, whether we point the negative (2pz) lobes both inward or both outward. Since the **GO** is taken to be positive, we adopt the convention that the (2pz) lobe closest to the **GO** takes the same sign as the **GO**. Of course we could have reversed all the orbital signs in Figure 3-4. In fact we could even have used opposite signs for the central **GO** and the pair of nitrogen orbitals without destroying the symmetry equality of the **GO** and the nitrogen (1s) orbitals. We adopt the convention, however, that the (1s) **GO** and the associated (1s) AOs on the atoms always have a positive sign.

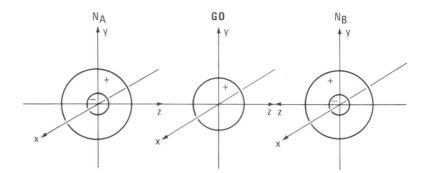

Figure 3-6. A depiction of the identity of the symmetries of a (1s) **GO** and a pair of (2s) AOs of the diatomic molecule $N_2$ in the field of the two nuclei.

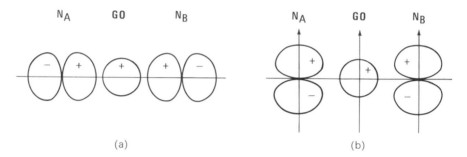

Figure 3-7. A depiction of the identity of the symmetries of a (1s) **GO** and a pair of (2pz) AOs of a diatomic molecule $N_2$ in the field of the two nuclei (a). In (b) it is seen that the symmetries of the (1s) **GO** and a pair of (2py) AOs are not identical in a diatomic potential.

Up to now we have seen nitrogen AO arrangements which are *permitted* by the (1s) **GO**. In other words, these permitted AO arrangements have the same symmetry in the diatomic potential as the **GO**. In Figure 3-7b we see a nitrogen AO arrangement which is *not permitted* by a (1s) **GO**, since no matter how we arrange the signs in the (2py) AOs, their symmetry in the diatomic potential is not the same as that of the (1s) **GO**. This can be seen by imagining a point $q$ anywhere in this combination of the (2py) AOs and inverting it through the molecular center. The sign of the wave function at $q'$ is seen to be opposite sign to that at $q$. While it is true that rotating the (2py) AOs in Figure 3-7b by 180° about the $z$ axis does not change the wave function, note that now reflection of a point $q$ through the $xz$ plane reverses the sign of the wave function. Another way of saying the same thing is that this combination of the (2py) AOs possesses a node in the $xz$ plane, and the (1s) **GO** does not have such a node. Therefore the symmetries of the nitrogen (2py) AOs and the (1s) **GO** are not compatible.

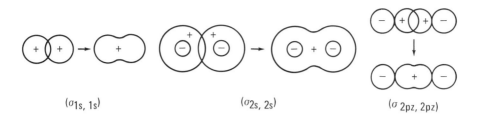

$(\sigma_{1s,\,1s})$                          $(\sigma_{2s,\,2s})$                          $(\sigma\,_{2pz,\,2pz})$

Figure 3-8. AO interactions in $N_2$ to form $\sigma$ molecular orbitals.

From the three *permitted* nitrogen AO arrangements we have generated thus far, we can form three MOs that can be written as linear combinations of atomic orbitals (LCAOs):

$$(\sigma_{1s,\,1s}) = N_1[(1sA) + (1sB)]$$

$$(\sigma_{2s,\,2s}) = N_2[(2sA) + (2sB)] \qquad (3\text{-}2)$$

$$(\sigma_{2pz,\,2pz}) = N_3[(2pzA) + (2pzB)].$$

Although it is not apparent from Figures 3-4–3-7, the nitrogen AOs are sufficiently close to interact with each other, and the algebraic addition of their contours implied in Equation 3-2 is depicted in Figure 3-8. LCAOs of this type are called *delocalized MOs* because the electron density contours are free to encompass more than one nucleus (here, two nuclei). Equally valid for the description of the bonding in diatomics are *localized* MOs which, as we will see later, can be obtained by hybridization of the delocalized MOs. From Figure 3-8 it is seen that the delocalized MOs described by Equation 3-2 are of the $\sigma$ type because their constituent canonical AOs have $m = 0$. It may be noted here that the advantages of directing the $z$ axes as shown in the coordinate system we have adopted is that the $(\sigma_{2pz,\,2pz})$ MO can be written as a sum (Equation 3-2) as can the $(\sigma_{1s,\,1s})$ and $(\sigma_{2s,\,2s})$ MOs.

Having considered the symmetry effect of the $(1s)$ **GO** on all the available nitrogen AOs, we find that we still must generate seven more MOs. To do this, we progress to the next nodally more complex **GO** which is of course the $(2s)$ **GO**. However, the $(2s)$ has the same symmetry in the diatomic potential as a $(1s)$ **GO** since it differs from the $(1s)$ only by the presence of a spherical node. It is *never* necessary to use a **GO** containing a spherical node, since the presence of such a node does not change the symmetry properties of a **GO** in the potential of a molecule, *no matter what its geometry*. Thus we skip the $(2s)$ **GO** and move on to the $(2p)$ **GO** set.

Arbitrarily beginning with a $(2pz)$ **GO**, we see from Figure 3-9 that only three AO arrangements are permitted by the $(2pz)$ **GO**. Note that in obtaining these AO arrangements, we use the convention discussed earlier that *the sign of the AO lobe nearest to a* **GO** *lobe is the same as the sign of that* **GO** *lobe*. If the AO has a node (as in Figure 3-9b and c), then as we cross that node from the

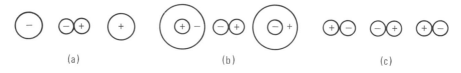

(a)                          (b)                          (c)

Figure 3-9. AO arrangements in $N_2$ permitted by the $(2pz)$ **GO**.

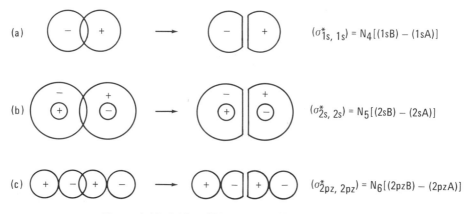

(a)        $(\sigma^*_{1s, 1s}) = N_4[(1sB) - (1sA)]$

(b)        $(\sigma^*_{2s, 2s}) = N_5[(2sB) - (2sA)]$

(c)        $(\sigma^*_{2pz, 2pz}) = N_6[(2pzB) - (2pzA)]$

Figure 3-10. MOs of $N_2$ generated by the $(2pz)$ **GO**.

lobe nearest to the **GO** to one further away, we of course change the sign of the amplitude.

The three AO arrangements in Figure 3-9 become MOs as shown in Figure 3-10a–c. A noteworthy feature of these three MOs is that, in contrast to the previous three, they contain a central planar node perpendicular to the internuclear axis because algebraic differences of contours are taken. This node is, of course, a consequence of the planar node in the $(2pz)$ **GO**. As we will come to realize more fully later, an MO without an internuclear node is a *bonding* MO (BMO) and an MO possessing such a node is an *antibonding* MO (ABMO). The latter type is designated by an asterisk, as shown in Figure 3-10.

Moving on to the $(2px)$ and $(2py)$ **GO**s, we see from Figure 3-11a and b that the symmetry of these GOs allows them to generate AO arrangements only among the $(2py)$ and $(2px)$ nitrogen AOs. These AO arrangements become the BMOs shown in Figure 3-11c and d. Since these MOs are generated by a pair of $(2p)$ **GO**s which are energetically degenerate in the diatomic potential, the MOs are also degenerate and thus have identical normalization coefficients in their expressions. It should be realized that these are BMOs since they contain no internuclear node *perpendicular* to the internuclear axis.

The remaining two MOs can be generated by a pair of $(3d)$ **GO**s, as shown in Figure 3-12. Having now generated all ten MOs for $N_2$, we can ask what effect the remaining $(3d)$ **GO**s and also the higher **GO**s have on the nitrogen AOs. It is always true that *once all the permitted AO arrangements are generated*

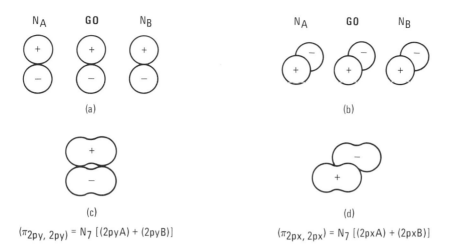

Figure 3-11. AO arrangements [(a), (b)] and MOs [(c), (d)] generated by the (2py) and (2px) **GO**s, respectively, in $N_2$.

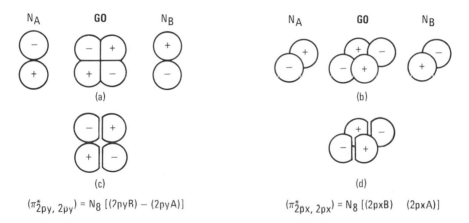

Figure 3-12. AO arrangements [(a), (b)] and MOs [(c), (d)] generated by the (3dyz) and (3dxz) **GO**s, respectively, in $N_2$.

by **GO**s, *all other **GO**s either generate no permitted AO arrangements or they reproduce permitted arrangements already obtained with the simpler **GO**s.*

Expressions for the normalization constants $N_1$ to $N_8$ in our diatomic nitrogen MOs can be obtained by realizing that these constants allow the integral over all space occupied by the electron density to equal 1. Thus in the case of $(\sigma_{1s, 1s})$, for example:

$$\int dV (\sigma_{1s, 1s})^2 = N_1{}^2 \left[ \int dV (1sA)^2 + \int dV (1sB)^2 + 2 \int dV (1sA)(1sB) \right]$$

$$(3\text{-}3)$$

which simplifies to Equation 3-4 when we recall that the integral of the square of an orbital equals 1:

$$N_1{}^2 \left[ 1 + 1 + 2 \int dV (1sA)(1sB) \right] = 1. \tag{3-4}$$

The integral $\int dV (1sA)(1sB)$ is a measure of the overlap between $(1sA)$ and $(1sB)$ and is called the overlap integral $S$. Thus we have $N_1 = [2(1 + S)]^{-1/2}$. There are mathematical techniques for evaluating overlap integrals but such calculations are beyond the scope of this book.

The qualitative LCAO treatment we have been developing permits us to estimate the positions of the MO energy levels relative to each other in most instances. There are mathematical models which yield more accurate expressions for MOs. These expressions often contain substantial contributions from more than two AOs in each MO, however. Although the LCAO approach is only an approximation, it is a useful one in most cases. Earlier we saw that the orbital energies for all atoms [except hydrogen for which $\varepsilon(2s) = \varepsilon(2p)$] are in the order $\varepsilon(1s) < \varepsilon(2s) < \varepsilon(2p)$. It is reasonable then that the energies of the MOs formed from these AOs lie in the order $\varepsilon(\sigma_{1s,\,1s}) < \varepsilon(\sigma_{2s,\,2s}) < \varepsilon(\sigma_{2pz,\,2pz})$. What about the relative positions of the BMO and ABMO formed from the same AOs, for example, $\varepsilon(\sigma_{1s,\,1s})$ versus $\varepsilon(\sigma_{1s,\,1s}^*)$? Although both have about the same electron density very close to the nuclei, the ABMO has a nodal plane at its center perpendicular to the internuclear axis. As we have seen before with AOs, the orbital energy rises as the number of nodes increases, and as expected $\varepsilon(\sigma_{1s,\,1s}) < \varepsilon(\sigma_{1s,\,1s}^*)$. Moving now to the MOs involving the nitrogen $(2p)$ AOs, we conclude from the above discussion that

$$\varepsilon(\pi_{2px,\,2px}) = \varepsilon(\pi_{2py,\,2py}) < \varepsilon(\pi_{2px,\,2px}^*) = \varepsilon(\pi_{2py,\,2py}^*). \tag{3-5}$$

Furthermore, we would expect that because the overlap integral $S$ in a $\pi$ or $\pi^*$ MO is smaller than in a $\sigma$ or $\sigma^*$ MO constructed from $p$ orbitals in the same set (Figure 3-13), the energy difference would be larger between $\varepsilon(\sigma_{2pz,\,2pz})$ and $\varepsilon(\sigma_{2pz,\,2pz}^*)$ than between $\varepsilon(\pi_{2py,\,2py})$ and $\varepsilon(\pi_{2py,\,2py}^*)$. Thus better overlap in a

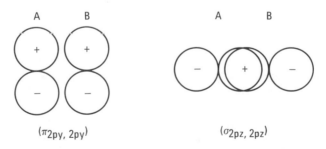

$(\pi_{2py},\,2py)$ $\qquad\qquad\qquad$ $(\sigma_{2pz},\,2pz)$

Figure 3-13. The smaller orbital overlap of a $\pi$ BMO compared to a $\sigma$ BMO constructed from $(p)$ AOs having the same principal quantum number.

BMO leads to stronger bonding (but only up to a point, as we shall see later) and also to stronger antibonding. We would expect then that the order of energies is as shown in Figure 3-14 wherein abbreviations for the MOs have been employed. There is evidence, however, that for most second-row diatomics in their ground states ($Li_2$ through $N_2$) the ($\sigma_p$) orbital lies *above* the ($\pi_p$), as shown in Figure 3-15a. When nuclear charges are low, the (2s) and (2p) AOs lie closer in energy. (Recall that in hydrogen these energies are actually the same.) Thus it is conceivable that considerable mixing of the (2s) and (2pz) orbitals can occur so that the resulting $\sigma$-type MOs are no longer of pure (2p) or of pure (2s) character. As we have seen earlier, such mixing of (2s) with (2p) on the atoms is expected to produce (sp) hybrids pointing in opposite directions on the z axis of each atom of nitrogen. To a first approximation, each hybrid on an atom could be thought of as an ($s^1p^1$) hybrid, although this is probably not the case for honomuclear diatomics since the chemical environment between the atoms is not the same as at the outside ends of the molecule. The consequence of such mixing is that ($\sigma_{2s}$) and ($\sigma_{2s}^*$) are stabilized (decrease in energy) and ($\sigma_p$) and ($\sigma_p^*$) are destabilized, as illustrated in Figure 3-15b. Thus the bracketed MOs in Figure 3-15 split apart

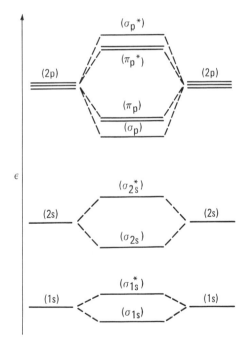

Figure 3-14. Energy level diagram for a diatomic molecule, showing the interaction of the atomic orbitals on nuclei A and B to form bonding and antibonding molecular orbitals.

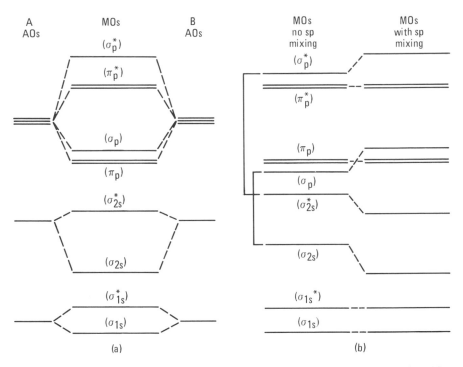

Figure 3-15. MOs of a diatomic molecule when there is $s-p$ mixing (a). In (b) the MOs which split apart under $s-p$ mixing are identified by connecting brackets.

because each bracketed pair is similar in energy and they have the same symmetries in the diatomic potential. Going across the Periodic Chart from left to right this splitting decreases because the separation of the valence $(2s)$ and $(2p)$ AOs increases. As a consequence of this process, $(\sigma_p)$ apparently does not drop in energy below $(\pi_p)$ for $O_2$ and $F_2$ (Figure 3-14).

It should be apparent that we can use **GOs** to generate MOs from the inward- and outward-pointing pairs of hybrids on $N_A$ and $N_B$ since they behave like $(2pz)$ AOs as far as their symmetry is concerned. This process is the subject of a problem at the end of the chapter.

Because the **GO** approach to bonding will be consistently employed throughout the remaining chapters, a summary of the rules developed so far governing the use of **GOs** is now given:

1.  Recognize that the number of valence AOs (VAOs) equals the number of MOs to be formed. Although we included the core $(1s)$ AOs in second-row diatomics, their overlap is relatively inconsequential because of their small size compared to the $(2s)$ and $(2p)$ AOs. Henceforth we will generally ignore MO formation from core AOs.
2.  Determine the permitted VAO combinations by placing first the $(1s)$, then the $(2p)$, $(3d)$, etc., **GOs** at the center of the molecule and then matching the signs of their wave functions in nearest lobes of VAO sets composed of the

*same* VAOs from one VAO set at a time. Thus only AOs of the same symmetry combine to form MOs [e.g., (2*px*A) combines with (2*px*B) but not with (2*py*B) or (2*s*B)]. Also, only AOs having similar energies in the free atoms combine to form MOs [e.g., (2*s*A) forms an MO with (2*s*B) but not with (1*s*B)]. When the permitted number of AO combinations has been obtained, the generation process is complete.

3. Formulate the LCAOs from the permitted VAO combinations as algebraic sums with a normalizing coefficient. Identify the MOs according to their $\sigma$ or $\pi$ nature by the absence or presence, respectively, of a planar node containing the internuclear axis, and determine whether or not the MOs are bonding or antibonding by the absence or presence, respectively, of a planar node perpendicular to the internuclear axis between the atoms.

## 3.3. Generator Orbitals and Molecular Motions

Above zero degrees Kelvin all molecules undergo translational and rotational motions in space, as well as vibrate internally. The translational and rotational motions can be thought of as having Cartesian components. Thus all molecules have three translational degrees of freedom and can therefore move along the $x$, $y$, and $z$ directions (or some linear combination of these directions). Similarly, all molecules, except linear ones, also possess three rotational degrees of freedom. Linear molecules have only two rotational degrees of freedom because rotations around the linear axis cannot be monitored. The remaining degrees of freedom belonging to molecules are vibrational modes. Another approach to molecular motions is to think of a molecule as a set of $n$ *independent* atoms (i.e., no bonds). Each atom then has three degrees of freedom corresponding to translation of the center of mass. Now think of fastening the atoms together with bonds. The $3n$ translational degrees of freedom formerly available to the assembly of unconnected $n$ atoms become reclassified, since only the molecule as a whole can translate, which accounts for three out of the total $3n$ degrees of freedom. With bond formation connecting all the atoms together, only the molecule as a whole can rotate, thus accounting for another three degrees of freedom if it is nonlinear. The rest of the degrees of freedom (i.e., $3n - 5$ for linear and $3n - 6$ for all other molecules) must be of the vibrational type. As we will see, vibrations result either from alternately stretching and compressing bonds or bending them back and forth.

**GO**s provide a convenient way to visualize what the vibrational modes of molecules look like. As with translational and rotational degrees of freedom, a vibrational mode also has components along the Cartesian coordinates. Thus we can think of the components of such a motion as a vector directed in either the positive or negative direction along $x$, $y$, or $z$. Let us consider such possible motions for the atoms in a diatomic molecule to be directed in the positive direction along the axes as depicted for $N_2$ in Figure 3-16. We will refer to such motions of atoms as atomic vectors (AVs). We realize, of course, that any

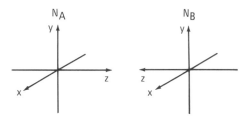

Figure 3-16. The axis systems chosen for the atomic vectors of the atoms in $N_2$.

atomic motion must reverse itself when its kinetic energy is expended in climbing the walls of the molecule's potential well by stretching and compressing the bond or by bending it. *All of the degrees of freedom (i.e., the translational plus rotational plus vibrational modes) are linear combinations of the atomic vectors (LCAVs) generated by* **GOs**.

Let us use $N_2$ as an example in generating the $3n - 5 = 1$ normal vibrational modes for any homonuclear diatomic molecule. We begin by noting that the *total* number of degrees of freedom is equal to the number of AVs, which in the present case is six. This parallels Rule 1 for generating MOs which says that the total number of VAOs equals the number of MOs. Next we parallel Rule 2 for MOs and place a $(1s)$ **GO** in the center of the molecule with sets of AVs of the same symmetry type, as shown in Figure 3-17a–c. *Vectors along the Cartesian axes behave as $(p)$ orbitals.* That is, $(p)$ orbitals have vectorial directionality in that they possess a positive and negative end and so can be represented by an arrow. It is therefore easily seen that Figure 3-17a represents a permitted AV combination whereas no combination of vectorial directions for either Figure 3-17b or c will match the symmetry of the $(1s)$ **GO**. Moving on to the $(2pz)$ **GO**, we see in Figure 3-18 that a permitted combination is

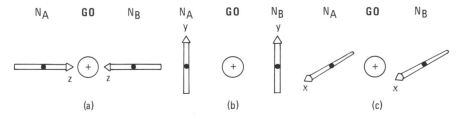

Figure 3-17. Atomic vector (AV) combinations in $N_2$ permitted (a) and not permitted [(b), (c)] by a $(1s)$ **GO**.

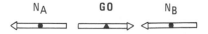

Figure 3-18. AV combination in $N_2$ generated by a $(2pz)$ **GO**.

generated with the $z$ atomic vectors. As with the placement of wave function signs in ($p$) orbitals, the orientation of the vectors is arbitrary. However, for symmetry reasons, no permitted combination can be generated by the ($2pz$) **GO** with the $y$ and $x$ AVs. Having generated two permitted AV combinations, we look to higher **GOs** to generate the remaining four shown in Figure 3-19. It may be noted in Figure 3-19c and d that the ($3d$) **GOs** are abbreviated with arrow drawings wherein the double-headed arrows denote the positive lobes and the doubly tailed arrows denote the negative lobes of the corresponding ($3d$) **GOs**. Instead of Rule 3 for MOs which involves writing normalized expressions, we will content ourselves with visualizing all the molecular motions of $N_2$ and identifying the vibrational mode. The motion depicted in Figure 3-17a is clearly the vibrational mode since both atoms are simultaneously moving toward the center. Of course, these atoms will move apart again, and we could generate that motion by employing a minus sign for the ($1s$) **GO** wave function. The motions in Figure 3-18 and in Figure 3-19a and b are translational modes of the molecule in the $z$, $y$, and $x$ directions, respectively. It is also easy to see that the motions represented by Figure 3-19c and d are rotational ones around the $x$ and $y$ axes, respectively. Note that no combination of AVs can represent a rotational motion around the $z$ axis for a linear molecule and indeed such a rotational motion does not exist.

It may appear that the **GO** approach to the bonding and molecular motion of diatomics is trivial. It should be realized, however, that the LCAOs and LCAVs for species such as tetrahedral $CH_4$ or octahedral $CoF_6^{3-}$ cannot be generated without **GOs** unless one happens to be well acquainted with group theory, a subject which actually turns out to be much easier to learn after one has gained an appreciation of **GOs**. As an aid to the three-dimensional visualization of the **GO** approach to molecular bonding and motion, the reader is referred to Appendix II wherein is described the construction of a set of simple models from inexpensive materials.

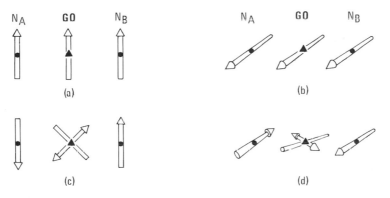

Figure 3-19. AV combinations in $N_2$ generated by the (a) ($2py$), (b) ($2px$), (c) ($3dyz$), and (d) ($3dxz$) **GO**.

## 3.4. Bond Strengths

A stretching vibration can be thought of as an oscillation of two masses separated by a spring with a force constant $k$ which characterizes the stiffness of the spring. The frequency $v$ of vibration is given by Equation 3-6:

$$v = 3906 \times 10^{10}[k(m_A^{-1} + m_B^{-1})]^{1/2} \text{ sec}^{-1} \qquad (3\text{-}6)$$

in which $m_A$ and $m_B$ are the atomic weights of the atoms in atomic mass units and $k$, the force constant, is in units of mdyne/Å. Values for $v$ range from $10^7$ to $10^8 \text{ sec}^{-1}$, which corresponds to infrared radiation. In general, strong bonds tend to be stiff (i.e., relatively large $k$ values) and the atoms tend to vibrate rapidly with small amplitudes. The converse is true for weak bonds.

Two critically important factors in determining the strength of a bond are the similarities of the energies of the neighboring-atom orbitals which can interact by virtue of their identical symmetries in the potential field, and the degree of overlap of these orbitals. Thus, for example, core ($1s$) electrons on atoms $N_A$ and $N_B$ of $N_2$ are at the same energy and are of the same symmetry in the diatomic potential, but they do not overlap nearly as strongly as the ($2s$) valence orbitals owing to the larger distance between the outer contours of the ($1s$) orbitals. The resultant $\sigma$ and $\sigma^*$ MOs are also split apart more in the case of the ($2sA$)–($2sB$) interactions than in that of the ($1sA$)–($1sB$) interactions.

The most widely used *quantitative* measure of the degree to which two orbitals overlap is the overlap integral $S$ discussed earlier $[S = \langle \phi | \psi \rangle = \int dV\, \phi(x, y, z)\psi(x, y, z)]$. Clearly, all non-zero contributions to $S$ come from those volume elements in which both orbitals have substantial non-zero values, i.e., where they overlap. The largest value of $S$ (which because of normalization equals 1) occurs when the orbitals are identical and superposed. $S$ equals zero if $\phi$ and $\psi$ are orthogonal. The overlap integral in molecules can have positive or negative values between $-1$ and $1$, depending on whether the overlapping orbital segments are of the same or opposite sign, respectively. Overlap integrals between wave functions $\phi$, $\psi$ centered on different nuclei in molecules never achieve the maximum value, i.e., $|1|$. To do so requires superposition of atoms which in turn requires coalescence of nuclei and electron clouds. Repulsive forces, of course, prevent nuclei from approaching each other very closely (except in nuclear fusion reactions). Why then does the equilibrium distance in molecules occur where it does? Some insight into this question can be gained by considering what happens to the energy splitting between two orbitals as the overlap changes (Figure 3-20). As the two orbitals approach one another, both their interaction energy and overlap integral increase but only up to a point. For reasons which need not concern us here, the energy splitting decreases after $|S|$ reaches ca. 0.5 to 0.65. It is now easy to see why we do not consider AOs of different symmetry when forming LCAOs in molecules. Thus, as shown in Figure 3-21, a ($2pzA$) and a ($2pyB$) have zero overlap.

It is instructive to observe that a plot of the potentials for the ($\sigma_{1s,\,1s}$)

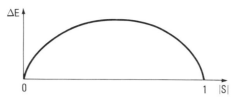

Figure 3-20. A plot of the energy splitting ($\Delta E$) between a BMO and its partner ABMO versus the absolute value of the overlap integral ($|S|$).

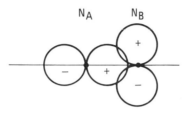

Figure 3-21. The net zero overlap of two ($2p$) orbitals of different symmetry in a diatomic potential.

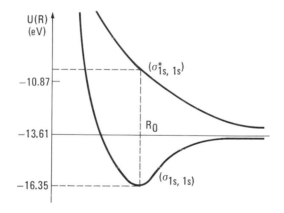

Figure 3-22. Plots of the potential energy of the BMO and ABMO of $H_2^+$ versus internuclear distance. For any given internuclear distance, the ABMO is more destabilized in energy than the BMO is stabilized.

orbitals of the $H_2^+$ ion (Figure 3-22) reveals that the ABMO has a *repulsive* potential. The energy $U(R) = 13.61$ eV in this figure represents $H_2^+$ dissociated into H and $H^+$. In other words, promotion of the electron in the ($\sigma_{1s, 1s}$) MO to the partner ABMO results in destabilization of the $H_2^+$ molecule and the component H and $H^+$ species fly apart. It is for this reason that we call the ($\sigma_{1s, 1s}^*$) an *anti*bonding MO. In fact, ABMOs in general are somewhat *more* antibonding than the corresponding BMOs are bonding (cf.

Figure 3-22). Thus each occupied BMO strengthens the bond and each occupied ABMO slightly more than offsets its equally occupied partner BMO. These considerations now allow us to define the concept of *bond order*. A single bond is afforded by double occupancy of a BMO (as in the ground state of $H_2$). Consequently, $H_2^+$ is held together by a half-bond. In general, we can define bond order by Equation 3-7.

$$\text{Bond Order} = \tfrac{1}{2}[(\text{Total No. e}^-\text{'s in BMOs}) - (\text{Total No. e}^-\text{'s in ABMOs})]$$
$$(3\text{-}7)$$

In Table 3-1 are collected the bond orders, dissociation energies, bond lengths, and force constants for some diatomic species. In general, dissociation energies and force constants increase and bond lengths decrease as bond orders rise.

## 3.5. Heteronuclear Diatomic Molecules

Unlike the MOs in homonuclear diatomics, the MOs in heteronuclear diatomics of necessity must be obtained by interacting nonidentical VAOs. This means interacting atomic VAOs of similar symmetry type but the *principal quantum numbers may be different*; in general, they are VAOs of different energies. It can also mean that VAOs of *different l quantum numbers* interact if they are sufficiently close in energy. Sometimes the choice of VAO pairs is not obvious, and it is further complicated by the relative degrees of VAO overlap, which can only be ascertained by complex calculations. When there

Table 3-1. Bond Strengths in Homonuclear Diatomic Molecules

| Molecule | Bond Order | Dissociation Energy (eV) | Bond Length (Å) | Force Constant (mdyne/Å) |
|---|---|---|---|---|
| $H_2^+$ | 0.5 | 2.74 | 1.06 | 1.4 |
| $H_2$ | 1 | 4.48 | 0.74 | 5.1 |
| $He_2^+$ | 0.5 | 3.1 | 1.08 | 3.1 |
| $He_2$ | 0 | no chemical bond | | |
| $Li_2$ | 1 | 1.03 | 2.67 | 0.25 |
| $Be_2$ | 0 | no chemical bond | | |
| $B_2$ | 1 | 3.0 | 1.63 | 3.5 |
| $C_2^+$ | 1.5 | 5.5 | 1.34 | |
| $C_2$ | 2 | 6.5 | 1.24 | 9.3 |
| $N_2^+$ | 2.5 | 8.72 | 1.12 | 19.5 |
| $N_2$ | 3 | 9.76 | 1.09 | 22.4 |
| $N_2^-$ | 2.5 | | | 13.2 |
| $O_2^+$ | 2.5 | 6.48 | 1.12 | 16 |
| $O_2$ | 2 | 5.12 | 1.21 | 11.4 |
| $O_2^-$ | 1.5 | | 1.28 | 5.6 |
| $F_2^+$ | 1.5 | 3.19 | 1.28 | |
| $F_2$ | 1 | 1.56 | 1.42 | 4.5 |
| $Ne_2$ | 0 | no chemical bond | | |

is strong competition between two VAOs on an atom, the MOs gain partial character of *both* contributing VAOs on that atom. For simplicity, we will assume that the VAOs closest in energy will interact as long as they have the proper symmetry. In Table 3-2 are some data which will be helpful to us in this regard.

Let us consider LiH as an example. The VAOs on Li include the ($2s$) and the ($2p$) set. Should we allow the (H$1s$) to interact with (Li$2s$) or with a (Li$2p$) pointed toward the (H$1s$) [i.e., (Li$2pz$)]? The energy of the Li ($2p$) AOs is not given in Table 3-2 because this table applies to the ground states of atoms and in the case of Li the ($2p$) level is unoccupied in the ground state. If we excite an electron from the ($2s$) to a ($2p$) orbital, the ($2p$) energy is about 8 eV. On the basis of relative energies, therefore, we would wish to interact the (H$1s$) orbital with a (Li$2pz$) orbital, although the difference in energy between the ($2s$) and ($2p$) levels in Li is relatively small compared to the gap between these levels in, say, boron through neon. Thus we might also expect some interaction between (H$1s$) and (Li$2s$). As we saw earlier, we must also consider the overlap

Table 3-2. Valence-Orbital Potential Energies[a] in eV

| Atom | $1s$ | $2s$ | $2p$ | $3s$ | $3p$ | $4s$ | $4p$ |
|------|------|------|------|------|------|------|------|
| H | 13.6 | | | | | | |
| He | 24.5 | | | | | | |
| Li | | 5.5 | | | | | |
| Be | | 9.30 | | | | | |
| B | | 14.0 | 8.3 | | | | |
| C | | 19.5 | 10.7 | | | | |
| N | | 25.5 | 13.1 | | | | |
| O | | 32.4 | 15.9 | | | | |
| F | | 46.4 | 18.7 | | | | |
| Ne | | 48.5 | 21.6 | | | | |
| Na | | | | 5.2 | | | |
| Mg | | | | 7.7 | | | |
| Al | | | | 11.3 | 6.0 | | |
| Si | | | | 15.0 | 7.8 | | |
| P | | | | 18.7 | 10 | | |
| S | | | | 20.7 | 12 | | |
| Cl | | | | 25.3 | 13.7 | | |
| Ar | | | | 29.3 | 15.9 | | |
| K | | | | | | 4.3 | |
| Ca | | | | | | 6.1 | |
| Zn | | | | | | 9.4 | |
| Ga | | | | | | 12.6 | 6.0 |
| Ge | | | | | | 15.6 | 7.6 |
| As | | | | | | 17.6 | 9.1 |
| Se | | | | | | 20.8 | 11.0 |
| Br | | | | | | 24.1 | 12.5 |
| Kr | | | | | | 27.5 | 14.3 |

[a] It should be recalled that these are negative energies since they represent attractive potentials between an electron and a shielded nuclear charge.

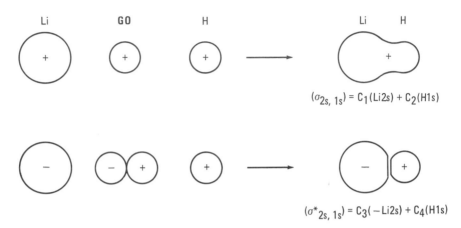

$(\sigma_{2s, 1s}) = C_1 (Li2s) + C_2 (H1s)$

$(\sigma^*_{2s, 1s}) = C_3 (-Li2s) + C_4 (H1s)$

Figure 3-23. The generation of the MOs in LiH from (H1$s$) and (Li2$s$) using **GO**s.

criterion. That is, how does the overlap between (H1$s$) and (Li2$pz$) compare to that between (H1$s$) and (Li2$s$)? From calculations it is known that the overlap is actually better between (H1$s$) and (Li2$s$). The resultant bonding in LiH according to this calculation involves a 64% contribution by (Li2$s$) and only a 36% contribution by (Li2$pz$). This mixture represents the compromise between energy and the overlap criteria. How, then, do we make a decision regarding orbitals to use in future examples? For reasons which will become clear later, this is not a severe problem.

Since the (Li2$s$) orbital is the main contributor to the bonding in LiH, let us, for the sake of simplicity, assume that it is the only orbital on lithium which interacts with (H1$s$). Note that in a *hetero*nuclear diatomic molecule, a point reflected across the $xy$ plane or inverted through the midpoint of the molecule does not feel the same potential regardless of which pair of orbitals we choose [including (Li2$s$) and (H1$s$)] because the nuclear charges are different. Thus we must be aware that we are dealing now with a molecule having less symmetry than a homonuclear diatomic. In Figure 3-23 we see how the two MOs arise from the generation of the VAO combinations permitted by the **GO**s. The corresponding MO energy level diagram is drawn in Figure 3-24. There are several important points to note from these figures:

1. Determination of the magnitude of the MO splittings is beyond the scope of this book. Thus we shall always be very qualitative about such splittings in our MO energy level sketches.
2. Notice that separate constants ($C_1$ through $C_4$) are used to describe contributions from the VAOs in the MO expressions. These constants are positive numbers which contain a normalization constant as well as a fraction representing the contribution of that particular VAO to the MO. Thus the BMO has more hydrogen (1$s$) character than lithium (2$s$) whereas

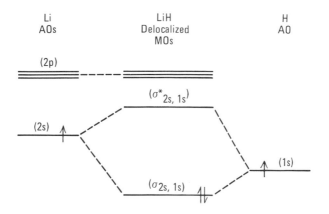

Figure 3-24. Delocalized MO energy level diagram for LiH.

the opposite is true for the ABMO because of the orthogonality require-ment. In other words, the BMO electron density is more concentrated near the H nucleus than the Li nucleus. A bonding electron excited to the ABMO conversely would spend most of its time near the Li nucleus.

3. The coefficients $C$ also depend upon one another since $\int dV(\psi)(\psi) = 1$. Thus the coefficients must accommodate the requirement that

$$\int dV(\sigma_{2s, 1s})^2 = \int [C_1(Li2s) + C_2(H1s)]^2 = 1. \qquad (3\text{-}7)$$

4. The electron pair bond in LiH is polar in that the electron density is shifted toward the hydrogen because its electronegativity exceeds that of lithium. Recall that the electronegativity of an atom reflects its tendency to with-draw electron density from neighboring atoms in a molecule. In homonu-clear diatomics the electronegativities are the same and so the bonds are nonpolar.

5. For our qualitative MO diagrams we will generally ignore the fact that ABMOs are somewhat more antibonding than the corresponding BMOs are bonding.

## 3.6. Localized MOs for Diatomics

In previous discussions we learned that the canonical delocalized orbitals in conjugated molecules such as butadiene could be transformed into localized orbitals having maximum amplitudes between carbon atoms. We also saw that the canonical AOs $(s)$, $(p)$, $(d)$, etc., could be recast into more localized hybrid sets $(sp)$, $(sp^2)$, etc. We will now show qualitatively that the delocalized diatomic canonical MOs can also be localized. The mathematical treatment will not be given here, but the results can be visualized from Figure 3-25. For

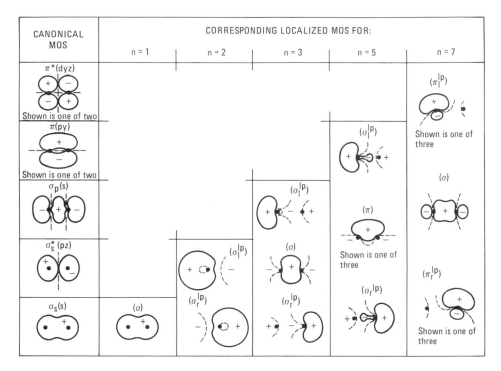

Figure 3-25. Chart of pictorial representations of canonical and corresponding localized MOs for homonuclear diatomics containing $n$ pairs of valence electrons (adapted from W. England, L. S. Salmon and K. Ruedenberg, *Topics in Current Chemistry*, **23**, 31 (1971)).

a molecule such as $Li_2$ which contains one pair of valence electrons ($n = 1$ in Figure 3-25) localization is meaningless and not possible. Thus the electron density is still concentrated between the two nuclei. For the hypothetical $Be_2$ molecule, $n = 2$ and we recall that the first two canonical MOs [$(\sigma_{2s, 2s})$ and $(\sigma^*_{2s, 2s})$] would have to be occupied. The two linear combinations of these two MOs obtained by superposing them are easily visualized to lead to $\sigma$-type orbitals which concentrate electron density along the internuclear axis but outside of the hypothetical molecule rather than between the nuclei. These are designated $\sigma_l^{lp}$ and $\sigma_r^{lp}$ where the subscripts stand for one orbital pointing to the left and the other to the right. Such electron pairs produce no bonding since there is no significant electron density between the nuclei. These electron pairs are associated largely with only one nucleus and are therefore called *lone pairs* (*lp*). The resulting lack of bonding between the two Be nuclei in the localized bonding view leads us to the same conclusion reached earlier with the delocalized MO picture, namely that $Be_2$ is not a stable molecule.

Before moving to $n = 3$, a word of clarification on lone pairs is in order. Lone pair orbitals are predominantly (*s*) in character since neutralization of

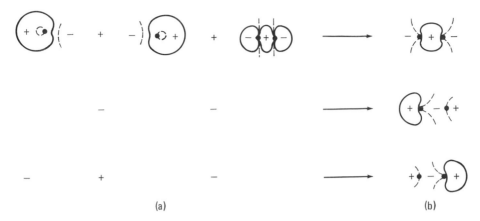

(a)                                                           (b)

Figure 3-26. Combinations of delocalized MOs (a) which give the localized MOs in (b).

the charge on one nucleus is more efficiently done by a spherical ($s$) orbital than by an orbital with directionality such as a ($p$). Bonding electron pairs tend to reside in orbitals having more ($p$) character since such AOs are more directional and bonding electrons are better able to neutralize the repulsive positive charges of two nuclei by residing in an MO made up of AOs which concentrate the electron density between the two nuclei. This becomes increasingly true for atoms toward the right of the Periodic Table for which the difference in energy between the valence ($s$) and ($p$) AOs increases (Table 3-2).

The molecule $B_2$ is an example of a system in which $n = 3$. However, from Figure 3-15, $B_2$ is seen to have two unpaired electrons. Difficulties arise with the localized approach when orbitals are singly occupied and we will therefore content ourselves with only the delocalized approach in such cases. Excitation of ground state $B_2$ does give rise to an excited state in which all the electrons are paired, however. Here we can form three localized MOs and the permitted linear combinations of delocalized MOs which produce them are shown in Figure 3-26.

It is interesting to note that in cases where there are lone pairs as well as bond pairs, the lone pairs generally lie higher in energy and in fact are more easily ionized. Indeed, the lone pair MOs are seen in Figures 3-25 and 3-26 to possess a node *between* the nuclei and hence they possess some *antibonding* character. What are the energies of the localized MOs for our excited $B_2$ molecule? Recall that the sum of the energies of the occupied orbitals must remain constant whether the orbitals are canonical or hybrids (Chapter 2.5). Thus the delocalized and localized MO energy level diagrams for the excited state of $B_2$ might look like those in Figure 3-27.

The ground state of $C_2$ ($n = 8$) has no unpaired electrons, and the visualization of its localized MOs is left as an exercise. Since the ground state of $O_2$ ($n = 12$) contains unpaired electrons, we will move on to the ten- and

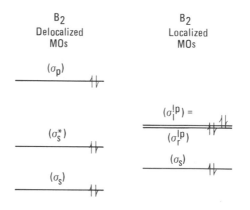

Figure 3-27. Delocalized and localized MO energy level diagrams for the lowest excited state of $B_2$.

fourteen-electron cases of $N_2$ and $F_2$, respectively. In the fifth and sixth columns of Figure 3-25 are the orthogonal localized MOs obtained by linearly combining the first five and all seven canonical MOs, respectively. Notice that $N_2$ is bonded by three equivalent "banana" bonds which are localized forms of the delocalized $\sigma_p$ and two $\pi_p$ MOs. We also observe that whenever a BMO–ABMO pair are both occupied in the delocalized bonding view (e.g., $N_2$ and $F_2$), we obtain a pair of *slightly antibonding* lone pair orbitals in the localized approach. The slightly antibonding nature of lone pair orbitals on adjacent atoms arises from the fact that delocalized ABMOs are more antibonding than the partner delocalized BMOs are bonding. The lone pair orbitals in $N_2$ and $F_2$ are shaped roughly like atomic ($sp$) and ($sp^3$) hybrids, respectively, and later we will return to this point.

From the localized view of diatomics we can see that lone pairs and bond pairs appear to repel one another. That is, in the excited state of $B_2$ each boron is surrounded by two electron pairs (one lone pair and one shared bond pair) and these pairs are concentrated in lobes on each atom which are 180° apart. In $F_2$ there are three lone pairs and a shared bond pair around each atom and they point roughly to the apices of a tetrahedron. That is, to the extent that the lone pairs have more ($s$) character, the angle they make with each other is $> 109°28'$ and their angle with the bond pair is $< 109°28'$. The idea of electron pair "repulsions" will be useful to us later in predicting molecular geometries.

*The localized view of diatomics (as well as of more complicated molecules, as we shall see) can also be generated from the delocalized view by a* **GO** *device similar to the one developed for the delocalized view.* Since for our purposes only fully occupied delocalized MOs can be localized, *we begin with Step 1 by identifying the* **GO**s *which give rise to the occupied delocalized MOs.* To simplify the process, *we do this after all BMO–ABMO pairs have been localized as lone pairs on the atoms.* Taking $F_2$ as an example, we see that after the latter procedure only one occupied MO is left and it is a BMO generated by an ($s$)

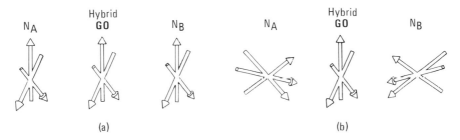

Figure 3-28. The use of **GO**s in $N_2$ to generate localized "banana" bonds composed of (*spxpy*) VAOs on each atom (a) and (*spxpy*) hybrids to which ($2pz$N) character has been added (b). The banana-like appearance of each bond is more apparent in (b) because of the angular overlap of the nitrogen hybrids.

**GO.** As we have already seen with $H_2$ (Figure 3-25), localization of one occupied MO does not alter a delocalized MO. Moving to $N_2$ as an example we note that an (*s*)-generated $\sigma$ and two (*p*)-generated $\pi$ BMOs remain after localizing the lowest-energy pair of MOs as lp's. The three BMOs are generated by (*s*), (*px*), and (*py*) **GO**s. *Step 2 is to hybridize the* **GO**s *which generate occupied delocalized MOs and use the* **GO** *hybrid set as a template at the molecular center to generate localized MOs.* It is seen from Figure 3-28a that the symmetry of such a **GO** set would "call in" the lobes of suitably oriented (*sp²*) VAOs on the nitrogens. Since we do not know the precise admixture of (*s*) and (*p*) character in the nitrogen hybrids, it is not difficult to see that by adding some (*pz*) character to the nitrogen (*sp²*) VAOs (Figure 3-28b), the banana bonds depicted in Figure 3-25 could arise through overlap of the main lobes of the nitrogen hybrid VAOs shown in Figure 3-28b. In fact, we indeed expect such an enrichment in (*pz*) character in these orbitals since, as we have seen earlier, the lp's on each nitrogen tend to have more (*s*) than (*p*) character and therefore the hybrids in which they are housed must have less than 50% (*pz*) character.

## Summary

After considering some of the factors leading to the formation of a diatomic molecule, the generator orbital device for visualizing delocalized MOs in diatomics was introduced and some rules developed for its application. It was seen that the order of delocalized MO energies in diatomics is determined by the degree of (*s*)–(*p*) mixing. The **GO** device was then applied to molecular motions in a diatomic molecule and rules parallel to those for **GO** generation of delocalized MOs were presented. The criteria of VAO energy and overlap as they relate to MO formation and bond strength were briefly examined and a definition of bond order was given. An approach to delocalized MOs in heteronuclear diatomics was then elaborated and some observations were made on characteristics of heteronuclear bonding. We concluded our discussion by developing rules for using **GO**s to generate localized MOs from delocalized ones.

PROBLEMS

1. Using sketches, show how the degeneracy of the ($3d$) **GO**s is broken when these AOs are at the center of a diatomic potential. Do the same for the ($4d$) AOs. Do the degeneracies of either set of orbitals break differently in a heteronuclear diatomic potential?

2. Show that two of the ($d$) **GO**s do not generate permitted AO arrangements for $N_2$. Which ($d$) **GO**(s) generates an MO already generated by a lower **GO**?

3. Using **GO**s, sketch the LCAOs for $N_2$ assuming the presence of inward-pointing $sp^{1.5}$ hybrid VAOs and outward-pointing $s^{1.5}p$ VAOs. Write expressions for all the MOs and draw an MO energy level diagram.

4. Using the **GO** approach, generate the permitted AO arrangements, formulate expressions for the delocalized MOs, sketch an MO energy diagram showing the appropriate orbital occupation, and give the overall bond order for $Rb_2$, $He_2^+$, BO, BeH, and diamagnetic $O_2$.

5. Of the species in Problem 4, which have nonpolar and which have polar bonds? For the latter, show the direction in which the electron density is polarized.

6. Account for the fact that removal or addition of an electron to $N_2$ weakens the bond.

7. Account for the fact that removal of an electron from $O_2$ strengthens the bond.

8. Account for the fact that the bonds in $N_2^-$, $O_2$, $O_2^-$, $F_2^+$, and $F_2$ are slightly weaker than the bonds of equal bond order in $B_2$, $C_2^+$, $C_2$, and $N_2^+$.

9. Considering the drawings of the localized MOs in Figure 3-25, sketch the localized MOs for ground state $C_2$. Using the **GO** device, generate the localized view of $C_2$ from the delocalized view and compare the results with the first part of this problem. Draw the delocalized MO energy level diagram for $C_2$ and right beside this diagram draw the corresponding diagram for the localized view you obtained. Describe the relative $s$ and $p$ character of all the localized orbitals in $C_2$.

# CHAPTER 4

# Linear Triatomic Molecules

Triatomic molecules are most frequently bent; less common are the linear and (the quite rare) triangular arrangements. Stable molecular geometries with precise bond angles and bond lengths can be predicted by determining quantum mechanically the molecular energy as a function of atomic positions and then finding the minima in this function. Sophisticated calculations are required for this purpose, however. For large molecules and compounds of heavier elements the computations are still too large even for modern computers to achieve substantial accuracy. Since one of our purposes is to develop *qualitative* delocalized and localized views of bonding in polyatomic molecules, we must first know their geometries. The concept most often used for this purpose is the so-called "*valence state electron pair repulsion*" or *VSEPR* approach developed by Gillespie and Nyholm, and indeed we shall in future chapters utilize this idea extensively. *In order to justify the use of this tool, however, we will in the first several examples assume that we know the molecular geometry.*

## 4.1. Linear FHF⁻

This stable anion is an example of *hydrogen bonding* in which the hydrogen is exactly midway between the two neighbor atoms. The hydrogen bonding in this ion is the strongest known (27 kcal/mole, whereas most hydrogen bonds are in the 1–10 kcal/mole range). We shall see that the exceptional stability of FHF⁻ can be understood in terms of chemical bonding.

For the delocalized view of this ion we begin by noting that we have a central atom between two peripheral atoms. Thus we have *two different* sets of atoms and we must examine *separately* the behavior of their VAO sets in the

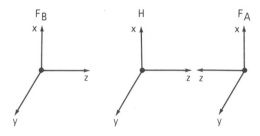

Figure 4-1. Axis system chosen for FHF⁻.

Table 4-1. Generator Table for FHF⁻

| VAO Equivalence Sets | GOs | | | | | |
|---|---|---|---|---|---|---|
| | $s$ | $pz$ | $px$ | $py$ | $dxz$ | $dyz$ |
| $H = (H1s)$ | n | | | | | |
| $Fs = (F_A2s), (F_B2s)$ | n | n | | | | |
| $Fp\sigma = (F_A2pz), (F_B2pz)$ | n | n | | | | |
| $Fp\pi = (F_A2px), (F_A2py), (F_B2px), (F_B2py)$ | | | n | n | n | n |

presence of **GO**s placed at the molecular center (i.e., at the hydrogen location) in the axis system shown in Figure 4-1. The VAOs available are those listed under the column headed VAO Equivalence Sets in Table 4-1. Each of the four sets are equivalence sets because for any point in the linear triatomic potential an equivalent point is generated by reflection through the $xy$ plane, by inversion through the center, and by rotation around the $z$ axis. In other words, VAOs within equivalence sets are those which are interchanged by a *symmetry operation*. In fact, the symmetry of the linear triatomic potential is seen to be the same as that of the diatomic potential. Each VAO equivalence set in Table 4-1 is denoted by a *set label* (i.e., H, Fs, Fp$\sigma$, Fp$\pi$) and an explicit listing of those VAOs which are equivalent in the potential of the molecule. The three equivalence sets arising from VAOs on each of the fluorines duplicate those we found for the $F_2$ molecule, and the symbols $\sigma$ and $\pi$ have the same meaning as for diatomic molecules. The **GO**s in Table 4-1 are denoted in abbreviated form and we will use this notation from now on. Taking first the $H = (H1s)$ equivalence set, we find that an $s$ **GO** generates the symmetry orbitals (SOs) in Equations 4-1–4-3 in the H, Fs, and Fp$\sigma$ equivalent sets:

$$\tilde{\sigma}(s|H) = (H1s) \tag{4-1}$$

$$\tilde{\sigma}(s|Fs) = N_1[(F_A2s) + (F_B2s)] \tag{4-2}$$

$$\tilde{\sigma}(s|Fp\sigma) = N_2[(F_A2pz) + (F_B2pz)]. \tag{4-3}$$

In these equations we use the symbol $\tilde{\sigma}(\mathbf{GO}|Equiv.\ Set)$ to denote a *symmetry orbital $\tilde{\sigma}$ which is generated by the specific **GO** in the indicated equivalence set*. Note that in the case of $\tilde{\sigma}(s|H)$ in Equation 4-1, the SO is the $(1s)$ AO on

Table 4-2. Generator Table for $F_2$

| VAO Equivalence Sets | GOs | | | | | |
| --- | --- | --- | --- | --- | --- | --- |
| | $s$ | $px$ | $py$ | $pz$ | $dxz$ | $dyz$ |
| $Fs = (F_A2s), (F_B2s)$ | b | | | a | | |
| $Fp\sigma = (F_A2pz), (F_B2pz)$ | b | | | a | | |
| $Fp\pi = (F_A2px), (F_A2py), (F_B2px), (F_B2py)$ | | b | b | | a | a |

hydrogen. *It is a general rule that the VAOs on central atoms in molecules are* SOs, because their nuclei coincide with the centers of the **GOs** which therefore also generate these **VAOs**. The remaining SOs appear in Equations 4-4–4-9.

$$\tilde{\sigma}(pz|Fs) = N_3[(F_A2s) - (F_B2s)] \tag{4-4}$$

$$\tilde{\sigma}(pz|Fp\sigma) = N_4[(F_A2pz) - (F_B2pz)] \tag{4-5}$$

$$\tilde{\pi}(px|Fp\pi) = N_5[(F_A2px) + (F_B2px)] \tag{4-6}$$

$$\tilde{\pi}(dxz|Fp\pi) = N_6[(F_A2px) - (F_B2px)] \tag{4-7}$$

$$\tilde{\pi}(py|Fp\pi) = N_5[(F_A2py) + (F_B2py)] \tag{4-8}$$

$$\tilde{\pi}(dyz|Fp\pi) = N_6[(F_A2py) - (F_B2py)]. \tag{4-9}$$

The symbol "n" in Table 4-1 means that the SOs we have generated are all nonbonding SOs. Recall that in the case of diatomics, the corresponding SOs are BMOs and ABMOs because the atoms are sufficiently close to each other to favor VAO overlap. Thus we can write Table 4-2 as a generator table for $F_2$. Because none of the fluorine SOs in Table 4-1 overlap significantly with each other, any electronic charge in these delocalized MOs would be confined to the fluorine atoms and so these MOs are nonbonding as indicated by the symbol "n" in Table 4-1. The SO on hydrogen is also nonbonding since it is composed of an orbital on only one atom anyway. Since we are neglecting overlap between non-neighboring atoms, the normalization constants $N$ are all equal to $2^{-1/2}$.

If all the SOs in Table 4-1 are nonbonding, how then do we account for the stability of FHF⁻? Recall from our discussion of diatomics that MOs of the same symmetry can interact (split) if they are sufficiently close in energy. Thus, for example, in $B_2$, linearly combining the two $s$-generated SOs (see Table 4-2 and substitute B for F) drives the $Bp\sigma$ BMO above the $Fp\pi$ BMOs in energy. Similarly, the $pz$-generated SOs split the ABMOs of the B$s$ and B$p\sigma$ equivalence sets apart. In Table 4-1 we have a set of two SOs generated by $pz$. Linearly combining these will not split them enough to gain a strongly bonding MO since these SOs involve non-neighboring atoms. However, we also have a set of three SOs generated by $s$. Linearly combining these will be profitable in terms of bonding because neighboring atoms are involved in the resulting MOs. The MOs can be visualized from Figure 4-2. In Figure 4-2a are shown the relative energies of the (H1$s$), (F2$pz$), and (F2$s$) VAOs. Since there is

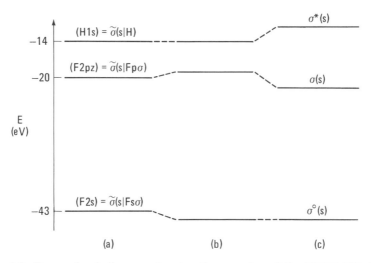

Figure 4-2. Energy level diagram showing the energies of the FHF⁻ SOs before interaction (a), after interaction of the two lowest-energy $s$-generated SOs (b), and after interaction of the two highest ones (c).

negligible overlap between VAOs *in the SOs*, it is clear that these VAO energies correspond in the first approximation to the SO energies. The relatively large separation of the two fluorine SOs of lowest energy causes their interaction to be small (Figure 4-2b) but the interaction between the two SOs closest in energy will be relatively large (Figure 4-2c), giving rise to a BMO $[\sigma(s)]$ and an ABMO $[\sigma^*(s)]$. The lowest-energy MO $\sigma^0(s)$ remains largely nonbonding and we denote this with the superscript shown. Expressions for these three MOs can be written as seen in Equations 4-10–4-12

$$\sigma^*(s) = a_1^* \tilde{\sigma}(s|Fs) + a_2^* \tilde{\sigma}(s|Fp\sigma) - a_3^* \tilde{\sigma}(s|H) \tag{4-10}$$

$$\sigma(s) = a_1 \tilde{\sigma}(s|Fs) + a_2 \tilde{\sigma}(s|Fp\sigma) + a_3 \tilde{\sigma}(s|H) \tag{4-11}$$

$$\sigma^0(s) = a_1^0 \tilde{\sigma}(s|Fs) + a_2^0 \tilde{\sigma}(s|Fp\sigma) + a_3^0 \tilde{\sigma}(s|H) \tag{4-12}$$

wherein $a_2^*$, $a_3^*$, $a_2, a_3$, and $a_1^0$ are relatively large positive constants causing the terms associated with them to be dominant contributors to their respective MOs. Confirmation of $\sigma^*(s)$ as an ABMO is realized from the presence of two nodes perpendicular to the molecular axis and lying between each pair of atoms. These nodes arise from domination of the last two terms in Equation 4-10. The $\sigma(s)$ MO has no internuclear nodes. Although $\sigma^0(s)$ also has no internuclear nodes, it may be noted that if we include the *internal* nodes of the VAOs there are zero, two, and four nodes in the $\sigma^0(s)$, $\sigma(s)$, and $\sigma^*(s)$ MOs, respectively, which is in accord with the rise in energy of these MOs in the same order. It should also be appreciated that $\sigma^0(s)$ is largely nonbonding mainly because $a_3^0$ is small.

Since the $a_1 \tilde{\sigma}(s|Fs)$ term is the main contributor to $\sigma^0(s)$ [owing to the low

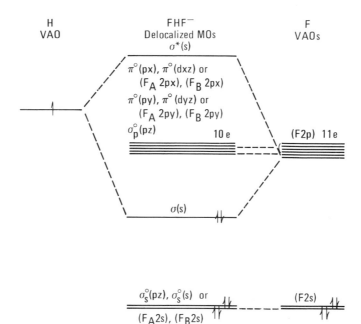

Figure 4-3. Delocalized MO energy level diagram for FHF⁻.

energy of $(F2s)$] we can for all intents and purposes write $\sigma^0(s)$ as Equation 4-2. At this point we also note that $pz$ generates an SO in the VAO equivalence set $Fs$, namely, Equation 4-4. As we saw in the case of diatomics having more than four electrons, the fully electron-occupied BMO–ABMO pair formed from *core* ($s$) VAOs can be recast into lone pairs on each atom. The same can be done for the two LCAOs in Equations 4-2 and 4-4 which, as we shall see shortly, are also occupied MOs. While there is also a splitting to be expected between the two $pz$-generated SOs in Equations 4-4 and 4-5, the interaction is expected to be small owing to the relatively large energy gap between the $(F2s)$ and $(F2p)$ VAOs. Thus $\tilde{\sigma}(pz|Fp\sigma)$, like $\tilde{\sigma}(pz|Fs)$, is a nonbonding MO and we give the former the label $\sigma_p^0(pz)$.

We are now ready to draw the orbital energy level diagram in Figure 4-3. First we note that there are sixteen valence electrons; one for H, seven each for the F atoms, and one for the overall negative charge (the latter electron having been arbitrarily placed in the F VAO set in Figure 4-3). These sixteen electrons occupy the lowest eight MOs. The lowest two MOs are atom-localized lp's. The next MO is a BMO $[\sigma(s)]$ *which is delocalized over three atomic centers.* Since it is the only BMO and it is doubly occupied, we have two electrons for both atom links, or an average bond order of $1/2$. Next in the diagram is a set of five *degenerate* fully occupied NBMOs which are devoid of contributions from the hydrogen. Each of these NBMOs is a 3-center 2-electron MO; four of them are of the $\pi$ type and one is of the $\sigma$ type. Although each of these NBMOs

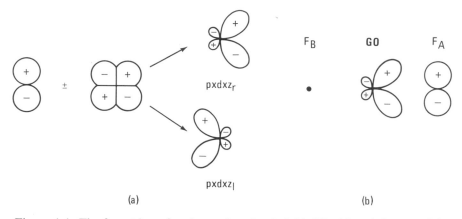

Figure 4-4. The formation of $pxdxz_r$ and $pxdxz_1$ hybrid **GO**s (a) and the use of the $pxdxz_r$ hybrid **GO** in calling in $(F_A 2px)$ (b).

is made up of only fluorine contributions, it is a mistake to think of them as representing an *unpaired* electron on each fluorine. Moreover, it has been shown experimentally that FHF$^-$ has no unpaired electrons. These NBMOs are simply orbitals with a relatively large degree of thinning out of the electron density in the region of the H atom lying between the main MO orbital lobes.

The localized view of FHF$^-$ is obtained by the same procedure as that used for diatomics. After first localizing the (F2s) electrons as a lp on each fluorine (see Figure 4-3), we see that there are six occupied MOs left; one bonding and the rest nonbonding. In $N_2$, we saw that localization of all BMOs at once (i.e., one $\sigma$ and two $\pi$-type occupied MOs) was convenient. Faced now with MOs which are of two types, namely, bonding and nonbonding, we will find it convenient to examine MOs in the same equivalence set in pairs. Let us begin with $\pi^0(px)$ and $\pi^0(dxz)$ generated from $(F_A 2px)$ and $(F_B 2px)$. Hybridization of the **GO**s which give rise to these occupied delocalized $\pi^0$ MOs leads to two $pxdxz$ **GO**s, as shown schematically in Figure 4-4a. Placing the $pxdxz_r$ **GO** at the center of the FHF$^-$ ion is seen to call in the $(F_A 2px)$ AO in Figure 4-4b. Similarly, $pxdxz_1$ calls in $(F_B 2px)$. Analogous examination of the delocalized $\pi^0(py)$, $\pi^0(dyz)$ pair leads to calling in of $(F_A 2py)$ and $(F_B 2py)$ via a pair of $pydyz$ **GO**s. Of course we can always localize still further the pair of occupied $(2p)$ orbitals called in on each fluorine by hybridizing them with the $(2s)$ lone pair on each of these atoms. By doing so we have an $sp^2$ hybrid set on each fluorine, as depicted in Figure 4-5a. Finally we consider the delocalized MO arising from the $Fp\sigma$ VAO set, namely $\sigma(s)$ and $\sigma_p^0(pz)$. The hybrid **GO** templates constructed from the $s$ and $pz$ **GO**s which generate $\sigma(s)$ and $\sigma_p^0(pz)$ yield the hybrid **GO**s $spz_r$ and $spz_1$, which are seen in Figure 4-5b to call in the $(2pz)$ VAOs on each fluorine. Because there is no $(pz)$ AO in the valence shell of hydrogen, however, we are unable to generate an $(spz)_r$ and an $(spz)_1$ VAO on the hydrogen. We are therefore left with only the (H1s) as a contributing VAO from hydrogen in the two localized orbitals. If we did have a $(pz)$ VAO

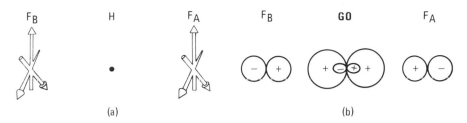

Figure 4-5. Hybrids of the ($sp^2$) type for fluorine lp's in FHF⁻. In (b) is shown the calling in of the fluorine ($2pz$) VAOs by the $spz_r$ and $spz_l$ **GO**s.

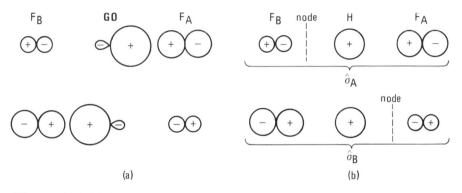

Figure 4-6. Orbital contributions from the fluorines in FHF⁻ generated by a pair of **GO**s (a) and the orbital contribution from hydrogen in this ion generated by these **GO** hybrids (b).

on a central atom (say X), we could have formed an occupied localized 2-center MO between $(Xspz)_r$ and $(F_A 2pz)$ and another between $(Xspz)_l$ and $(F_B 2pz)$. Examples of this type will be seen later. But how do we deal with the present case? It can be shown (Appendix III) that *when the central atom does not cooperate fully in responding to the hybrid **GO** template by supplying a matching set of hybrid VAOs, then pure 2-center localized MOs cannot be formed.* In other words, the MO formed is *mostly* localized as a 2-center BMO between the central atom and one of the peripheral atoms, but there is a *small antibonding* contribution between the central atom and the *other* peripheral atom. *This can be visualized (Figure 4-6) by recognizing that in such cases the sign of the VAO on the lesser contributing atom is chosen so as to produce a node between it and the VAO contributed by the central atom.* By this procedure we maintain the orthogonality of the orbitals. Note that there are no bonding–antibonding contributions in the MO in Figure 4-4b, for example, because it *does not* include a contribution from the central atom.

Since the two partially localized MOs which are denoted by $\hat{\sigma}_A$ and $\hat{\sigma}_B$ in Figure 4-6b result from equal mixing of a delocalized BMO and an NBMO (see also Appendix III), each localized MO must have a bond order of less

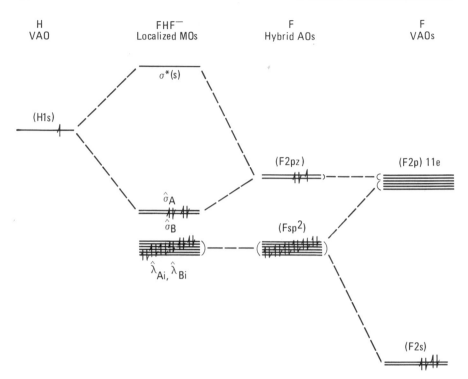

Figure 4-7. Localized MO energy level diagram for FHF⁻.

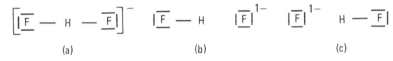

Figure 4-8. Electron dash structures for FHF⁻.

than 1. Thus each partially localized MO ($\hat{\sigma}_A$ and $\hat{\sigma}_B$) is concluded to have a bond order of 0.5 in accord with the delocalized view wherein the 3-center BMO has an average bond order of 0.5 per link. The energy level diagram for the localized bond picture is given in Figure 4-7. We should note that because of the large electronegativity difference between hydrogen and fluorine, the electron density in the BMO tends to concentrate on the fluorines, thereby giving rise to considerable *polarity (ionicity)* in the HF links.

The eight localized MOs can be represented by the familiar electron dash formulation in Figure 4-8a, provided we bear in mind that the two bond dashes are really only half-bonds. Usually, of course, we associate a dash with a full electron pair bond. Alternately, we can represent the bonding in FHF⁻ by the *average* of the conventional electron dash structures in Figure 4-8b and c. We see then that our conclusions regarding the localized view of FHF⁻ can be abbreviated by an electron dash structure in which six of the eight localized MOs are fluorine lp's. In fact, as a general rule, *whenever the number of occupied delocalized MOs equals the total number of VAOs in the VAO equivalence sets which make up these MOs, then the localized MOs obtained from these delocalized MOs are identical with the VAOs themselves*. This is true irrespective of the number of VAOs in the equivalence sets. This rule permits considerable simplification of our future discussions. In FHF⁻ we could thus start with the implication of the electron dash structure and *preassign* our lone pairs as fluorine canonical VAOs in the delocalized view and as hybrid VAOs in the localized view. In FHF⁻ the lp's are on non-neighboring atoms so that mutual overlap is negligible. Therefore, the orthogonalizing admixtures from the "other" atom are of little consequence and the lp's are for all intents and purposes purely nonbonding. Recall that in $F_2$ the lp's are on neighboring atoms, where overlap is more important, and correspondingly there are slight antibonding effects between lp's from neighboring atoms known as *"non- bonded" repulsions*. As a result of preassignment of lp's, only the $(F_B 2pz)$, $(H1s)$, and $(F_A 2pz)$ VAOs would be left to consider in the bonding treatments.

## 4.2. FXeF

Although procedures for drawing electron dash structures have not yet been discussed, let us adopt the one shown for FXeF in Figure 4-9. This electron dash structure indicates that we can localize three lp's on each atom. Using the same axis system as for FHF⁻ (Figure 4-1), each fluorine in the delocalized view would use its $(2s)$, $(2px)$, and $(2py)$ VAOs for lone pairs and the xenon would similarly utilize its $(5s)$, $(5px)$, and $(5py)$ VAOs. In the localized view, these lp VAOs could be hybridized to $(sp^2)$ VAOs. From our treatment of FHF⁻, it is clear that we then have only the $(F2pz)$ VAOs to consider for bonding if three lone pairs are allocated to each fluorine atom. In the valence shell of xenon we have a $(5pz)$ orbital as well as a set of $(5d)$ VAOs. Since the extent of $(d)$ orbital participation in $\sigma$ bonding is, in general, controversial, we will examine the bonding in FXeF both excluding and including an $(Xe5d)$ VAO. While $(d)$ orbitals are present in levels where $n$, the principal quantum number, exceeds 2, they lie sufficiently high in energy above the $(p)$ VAOs of the same principal quantum number that their inclusion as VAOs in nonmetals is quite questionable. Because of the shielding effect, however, $(d)$ orbitals do become important in bonding interactions in the transition metals.

For the delocalized view of FXeF *excluding* $(Xe5d)$ *participation* we can

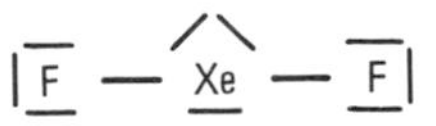

Figure 4-9. Electron dash structure for FXeF.

Table 4-3. Generator Table for
FXeF

| VAO Equivalence Sets | GOs | |
| | $s$ | $pz$ |
| --- | --- | --- |
| $\mathrm{Xe}p\sigma = (\mathrm{Xe}5pz)$ | | n |
| $\mathrm{F}p\sigma = (\mathrm{F_A}2pz), (\mathrm{F_B}2pz)$ | n | n |

write Generator Table 4-3. The three SOs generated by **GO**s are given in Equations 4-13–4-15:

$$\tilde{\sigma}(s|\mathrm{F}p\sigma) = N_1[(\mathrm{F_A}2pz) + (\mathrm{F_B}2pz)] \qquad (4\text{-}13)$$

$$\tilde{\sigma}(pz|\mathrm{F}p\sigma) = N_2[(\mathrm{F_A}2pz) - (\mathrm{F_B}2pz)] \qquad (4\text{-}14)$$

$$\tilde{\sigma}(pz|\mathrm{Xe}) = (\mathrm{Xe}5pz). \qquad (4\text{-}15)$$

The first SO, $\tilde{\sigma}(s,|\mathrm{F}p\sigma)$, must be an NBMO since it has no contribution from the Xe atom and because it is the only SO generated by the $s$ **GO**. Since the second and third SOs are generated by the same **GO**, they will interact to form a BMO and an ABMO. The delocalized MOs are thus

$$\sigma^0(s) = \tilde{\sigma}(s|\mathrm{F}p\sigma) \qquad (4\text{-}16)$$

$$\sigma(pz) = a_1\tilde{\sigma}(pz|\mathrm{Xe}) + a_2\tilde{\sigma}(pz|\mathrm{F}p\sigma) \qquad (4\text{-}17)$$

$$\sigma^*(pz) = a_1^*\tilde{\sigma}(pz|\mathrm{Xe}) - a_2^*\tilde{\sigma}(pz|\mathrm{F}p\sigma). \qquad (4\text{-}18)$$

After preassignment of the lp's we have four electrons left and so only the 3-center bonding and nonbonding MOs are occupied. The MO energy level diagram in Figure 4-10 indicates that again, as in FHF⁻, one electron pair binds the molecule in a (polar) 3-center bond and the other electron pair resides in an NBMO. The essential difference in the bonding between the two species is that an ($s$) VAO is used on the central atom in FHF⁻ while a ($pz$) VAO is used in FXeF.

For the localized view of FXeF [without Xe ($5d$) participation] we begin by localizing the fluorine lp's in ($sp^2$) fluorine VAOs and the xenon lp's in xenon ($sp^2$) VAOs. The remaining delocalized MOs are $\sigma(pz)$ and $\sigma^0(s)$ from which we obtain a pair of $spz_r$, $spz_1$ **GO**s with which to generate localized MOs in the same way as we did for FHF⁻. In the present case, however, only the (Xe5$pz$) VAO responds to the pair of hybrid **GO**s [since (Xe5$s$) contains a preassigned lp] and again the central atom VAO is involved in *both* partially localized MOs, as depicted in Figure 4-11b. Note in the SOs in Figure 4-11a and the MOs in Figure 4-11b that the ($2pz$) VAO of the fluorine atom providing the smaller contribution has the *opposite* sign as compared to its counterpart in Figure 4-6 for the case of FHF⁻. As pointed out earlier, this is to preserve orthogonality of the orbitals. In other words, to have some antibonding character in the MOs of Figure 4-11b, which now involve a ($pz$) VAO on the central atom, the signs of the VAOs must be as shown (see also Appendix III).

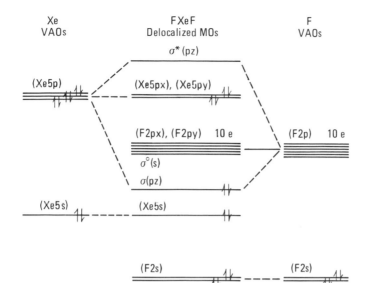

Figure 4-10. Delocalized MO energy level diagram for FXeF excluding xenon ($5d$) VAO participation.

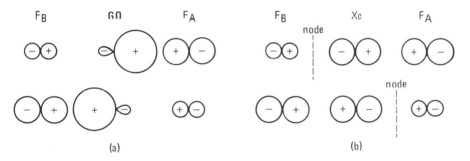

Figure 4-11. Orbital contributions from the fluorines in FXeF generated by a pair of *spz* **GO**s (a) and the orbital contribution from xenon in this molecule generated by these **GO** hybrids (b).

In FHF$^-$, the central atom provided an ($s$) VAO which requires the VAO signs to be as depicted in Figure 4-6.

In the localized MO energy diagram shown in Figure 4-12 we must recognize that the energy of the localized BMOs is the average of the localized $\sigma(pz)$ and $\sigma^0(s)$ MOs of Figure 4-10. This arises in the localized MOs $\sigma_A$ and $\sigma_B$ by virtue of their slightly antibonding nature which produces a bond order of 0.5 in each MO.

As with FHF$^-$, it is not possible to draw an electron dash structure for FXeF where each dash corresponds to a lp or a 2-center 2-electron bond. The

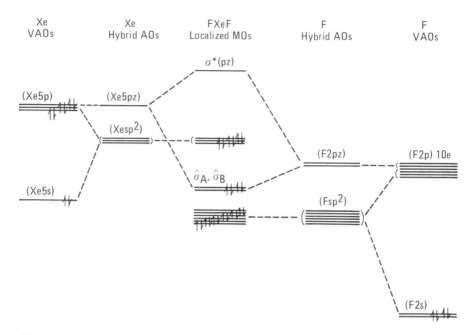

Figure 4-12. Localized MO energy level diagram for FXeF without xenon ($5d$) VAO participation.

Figure 4-13. Ionic electron dash structures for FXeF.

diagram in Figure 4-9 is deficient in that it implies the availability of five VAOs on xenon when we included only four. Furthermore, we saw that each of the two localized MOs $\sigma_A$ and $\sigma_B$ must use VAOs on *all three atoms* because *both* MOs must use one and the same VAO on the xenon and still be mutually orthogonal. As in FHF⁻, therefore, the bonding in FXeF must be symbolized by the average of the two ionic electron dash structures shown in Figure 4-13a and b.

What if we allow (Xe$5d$) participation in the VAO set of xenon? As shown in Figure 4-14, the only ($d$) orbital of the correct symmetry to interact with the fluorine SOs indicated in Table 4-3 is (Xe$5dz^2$). It is also apparent that the (Xe$5dz^2$) is an SO which is generated by $dz^2$ at the molecular center. Having now introduced a new **GO**, however, we must examine its effect on *all* the other equivalence sets in Generator Table 4-3. Clearly $dz^2$ generates exactly the same SO generated by the $s$ **GO** in the F$p\sigma$ equivalence set. Which **GO** shall we use then? Or shall we include both? Including both is unreasonable in that redundancy is thereby introduced. It makes sense to use that **GO** which

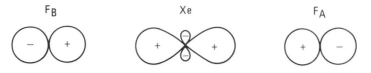

Figure 4-14. The generation by a $3dz^2$ **GO** of an SO in FXeF composed of a $(2pz)$ VAO on each fluorine. The $3dz^2$ **GO** also generates the $(Xe5dz^2)$ VAO, which is also an SO. Even though the latter AO contains two spherical nodes, its symmetry is the same as a $3dz^2$ **GO**.

Table 4-4. ´Generator Table for FXeF
[Including $(Xe5dz^2)$]

|  | GOs | |
| --- | --- | --- |
| VAO Equivalence Sets | $dz^2$ | $pz$ |
| $Xep\sigma = (Xe5pz)$ |  | n |
| $Xed\sigma = (Xe5dz^2)$ | n |  |
| $Fp\sigma = (F_A2pz), (F_B2pz)$ | n | n |

generates the most SOs and so *in circumstances like this we always utilize only the* **GO** *which is more inclusive.* Based on this reasoning the FXeF generator table assumes the form given in Table 4-4. Taking linear combinations of the SOs generated by the same **GO** leads to two BMO–ABMO pairs: one pair generated from $dz^2$ and the other from $pz$. The delocalized MO energy level diagram is shown in Figure 4-15.

The corresponding localized view is obtained by generating two localized BMOs from the delocalized ones via their $pz$ and $dz^2$ **GOs**. Taking linear combinations of these **GOs** leads to a pair of digonal hybrids $pzdz_l^2$ and $pzdz_r^2$, depicted in Figure 4-16. The directionality of these hybrids generates two localized MOs, one between Xe and $F_A$ and the other between Xe and $F_B$. The manner in which these **GOs** dictate this directionality is by generating two digonal $(Xepd)$ hybrids, each of which combines with the $(F2pz)$ VAO toward which it points to form a localized 2-center 2-electron MO concentrated in each Xe–F link. A schematic drawing of these localized bonds together with the $(sp^2)$-hybridized lp's is shown in Figure 4-17. The localized MO energy level diagram shown in Figure 4-18 tells us that the localized MOs $\hat{\sigma}_A$, $\hat{\sigma}_B$ are 2-center 2-electron fully bonding MOs (i.e., the bond order is 1.0) which lie at the average of the energies of the $\sigma(pz)$ and $\sigma(dz^2)$ delocalized MOs in Figure 4-15.

## 4.3. OCO

The electron dash structure normally drawn for this molecule is shown in Figure 4-19a. It implies that we have two lp's on each oxygen and a $\sigma$ plus a $\pi$ bond in each link. If we were to preassign an $(O2s)$ and an $(O2p)$ VAO on each

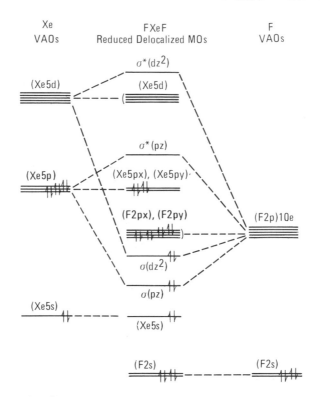

Figure 4-15. Delocalized MO energy level diagram for FXeF including $(Xe5dz^2)$ participation.

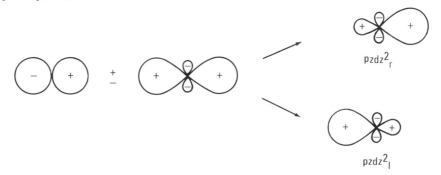

Figure 4-16. Formation of two $pzdz^2$ **GO** hybrids from $pz$ and $dz^2$ orbitals.

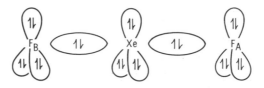

Figure 4-17. Localized lp's and bond pairs in FXeF, assuming $(Xe5dz^2)$ participation in the bonding.

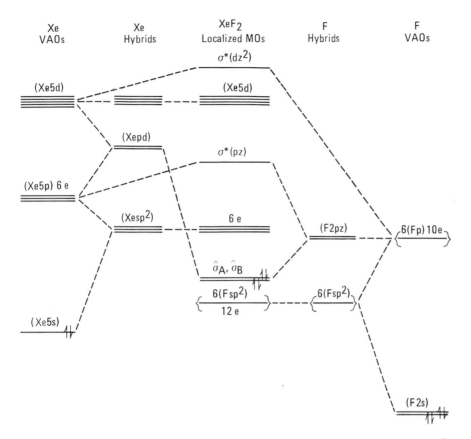

Figure 4-18. Localized MO energy level diagram for FXeF including $(Xe5dz^2)$ participation.

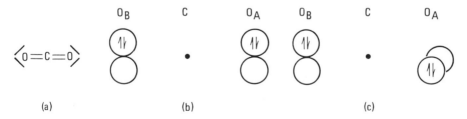

Figure 4-19. Electron dash structure for $CO_2$ (a) and two ways of preassigning its oxygen lp's [(b), (c)].

Table 4-5. Generator Table for $CO_2$

| VAO Equivalence Sets | GOs | | | | | |
|---|---|---|---|---|---|---|
| | $s$ | $pz$ | $px$ | $py$ | $dxz$ | $dyz$ |
| $Cs = (C2s)$ | n | | | | | |
| $Cp\sigma = (C2pz)$ | | n | | | | |
| $Cp\pi = (C2px), (C2py)$ | | | n | n | | |
| $Op\sigma = (O_A2pz), (O_B2pz)$ | n | n | | | | |
| $Op\pi = (O_A2px), (O_B2px), (O_A2py), (O_B2py)$ | | | n | n | n | n |

oxygen to accommodate these lp's, a problem with delocalization of the $\pi$ bonds arises. Thus if the ($p$) VAO used for a lp on each oxygen is $(O_A2px)$ and $(O_B2px)$ (Figure 4-19b) we are restricted to forming *one* 2-electron 3-center delocalized $\pi$ bond [i.e., with the $(2px)$ VAOs on all three atoms] when we really need two such delocalized $\pi$ bonds [i.e., an additional one involving all three $(2py)$ VAOs]. If we use $(O_B2px)$ and $(O_A2py)$ for one of the lp's on each O atom (Figure 4-19c) we can form no $\pi$ bonds delocalized over all three centers; only two 2-center 2-electron $\pi$ bonds [i.e., one in the $yz$ plane on the right side and one in the $xz$ plane on the left utilizing $(O_A2py)$, $(C2py)$ and $(O_B2px)$, $(C2px)$, respectively]. Since we wish to delocalize the bonding as much as possible, let us preassign only one lone pair on each oxygen to an $(O2s)$ VAO.

Using the axis system for $FHF^-$ in Figure 4-1 we can then write Generator Table 4-5, wherein ten VAOs form five equivalence sets. The ten SOs can be straightforwardly visualized and expressions written for them from the **GO** symmetries, as we have seen earlier. Since the $dxz$ and $dyz$ generate only one SO each, both SOs are MOs, and since they have no contributions from a neighboring atom, they are nonbonding:

$$\pi^0(dxz) = \tilde{\pi}(dxz|Op\pi)$$
$$\pi^0(dyz) = \tilde{\pi}(dyz|Op\pi). \tag{4-19}$$

The $px$ and $py$ **GO**s each generate two SOs: one on the central atom and one on the peripheral atoms. Their interaction gives rise to two BMO–ABMO pairs, namely, $\pi(px)$, $\pi^*(px)$ and $\pi(py)$, $\pi^*(py)$, with the two BMOs degenerate and the two ABMOs also degenerate. Next we consider the $s$- and $pz$-generated SOs, of which there are two in each set in Table 4-5. These interact to give BMO–ABMO pairs of the $\sigma$ type, namely, $\sigma(s)$, $\sigma^*(s)$ and $\sigma(pz)$, $\sigma^*(pz)$. There are sixteen electrons in the VAOs of which we have preassigned four to two $(O2s)$ lone pair orbitals. From Table 4-5 we generated two NBMOs, four BMOs, and four ABMOs. Clearly the twelve remaining electrons occupy all of these MOs except the four ABMOs, which are highest in energy. As seen in Figure 4-20, we have two occupied *3-center* $\sigma$ BMOs and two occupied *3-center* $\pi$ BMOs which together provide a bond order of 2.0 per link as expected from the electron dash structure. Notice, however, that our two occupied *3-center* $\pi$ NBMOs are *not* implied by the electron dash structure. This situation reminds us of FXeF in which one of the electron dashes

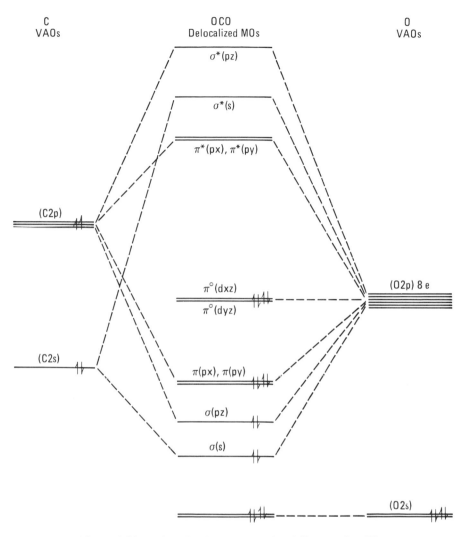

Figure 4-20. Delocalized MO energy level diagram for $CO_2$.

drawn between a pair of atoms had to be understood to symbolize a delocalized nonbonded electron pair. In $CO_2$ one of the oxygen lone pairs on *each* oxygen must be viewed as residing in a π NBMO in the delocalized view. These π NBMOs are *3-center* 2 electron MOs which lie at right angles to one another.

To generate the corresponding localized view, let us localize the two σ and the two π occupied BMOs in Figure 4-20 separately. The σ BMOs are generated by *s* and *pz* and so we place a pair of oppositely directed *spz* hybrid **GO**s at the molecular center as shown in Figure 4-21a. Each of these **GO**s leads to a localized BMO shown in Figure 4-21b confined to two neighboring

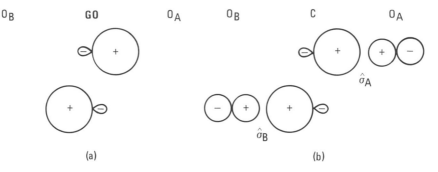

Figure 4-21. The *spz* hybrid **GO**s (a) and the localized MOs they generate (b) in $CO_2$.

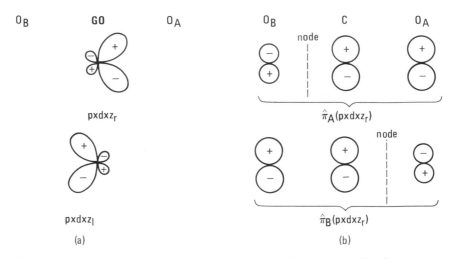

Figure 4-22. The *pxdxz* hybrid **GO**s (a) and the partially localized MOs they generate (b) in $CO_2$.

atoms with *no* contribution from the third atom. This assumes that the hybrid VAO on carbon is a *pure* (*sp*) hybrid. Actually such a hybrid will probably be $sp^\alpha$, where $\alpha$ is not quite 1. In that case the localized BMOs $\hat{\sigma}_A$ and $\hat{\sigma}_B$ will, of course, have to contain small antibonding contributions from $(O_B 2pz)$ and $(O_A 2pz)$, respectively, in order to preserve orthogonality.

Turning now to localization of the occupied $\pi$ and $\pi^0$ MOs, their **GO**s are hybridized to give the $pxdxz_1$, $pxdxz_r$ and $pydyz_1$, $pydyz_r$ **GO**s discussed earlier in connection with the $\pi^0$ MOs of $FHF^-$. The **GO**s in Figure 4-22a are seen in Figure 4-22b to call in only the (2px) VAO on carbon since no (*d*) VAOs are available on this atom. Hence there must be a small negative contribution from one of the oxygens, in addition to the substantial localization of bonding electron density in the opposite CO link stemming from the positive contribution of the other oxygen. These arguments are also applied to the localiza-

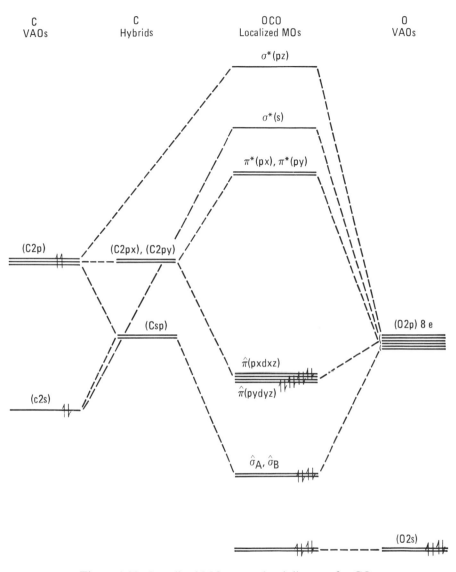

Figure 4-23. Localized MO energy level diagram for $CO_2$.

tion of the occupied $\pi(py)$ and $\pi^0(py)$ MOs. Because of the antibonding contributions in $\hat{\pi}_A(pxdxz)$, $\hat{\pi}_B(pxdxz)$, $\hat{\pi}_A(pydyz)$, and $\hat{\pi}_B(pydyz)$, these localized MOs are only half-bonding between each link. Thus each link contains one full $\sigma$ bond and two $\pi$ half-bonds, yielding a total bond order of $1 + 0.5 + 0.5 = 2.0$. The MO energy level diagram for this localized view is shown in Figure 4-23.

A second localized view of $CO_2$ is obtained by equivalent hybridization of the **GO**s which generate *all* the *bonding* MOs, namely, $sp^3$ hybridization. This

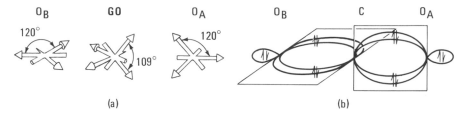

(a)                                                    (b)

Figure 4-24. Oxygen VAO hybrids called in by $sp^3$ **GO**s for localized bonds (a) and the resulting localized banana bonds (b) in $CO_2$.

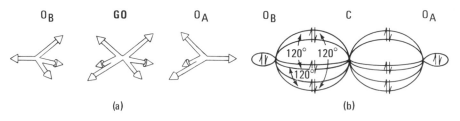

(a)                                                    (b)

Figure 4-25. Oxygen VAO hybrids called in by $sp^3dxzdyz$ **GO** hybrids (a) and the resulting banana bonds in $CO_2$ (b). Note that only the main lobes of all the hybrid AOs are shown in (a). This procedure will also be followed in future diagrams.

procedure gives a tetrahedral **GO** hybrid set which calls in two lobes of an $sp^2$ VAO hybrid on each oxygen, as shown in Figure 4-24a. [Here each of the (O2s) lp's has been hybridized into the $(sp^2)$ VAO set.] As seen in Figure 4-24b, this procedure leads to two pairs of mutually perpendicular banana bonds. The $\pi^0(dxz)$, $\pi^0(dyz)$ MOs now *cannot* be localized because hybridizing the $dxz$, $dyz$ **GO**s merely gives us another identical VAO set rotated by some angle. Notice that the bond order from Figure 4-24 is still 2.0 as expected.

What if we were to hybridize the **GO**s which generated *all six* of the occupied MOs in Figure 4-20? The hybrid **GO** set is then a trigonal prismatically disposed $sp^3d^2$ set which, as shown in Figure 4-25a, calls in three of the lobes of an $sp^3$ VAO set on each oxygen atom. Notice that an $(sp^3dxzdyz)$ hybrid set has a trigonal prismatic geometry whereas an $(sp^3dx^2 - y^2dz^2)$ hybrid set is octahedral. In Figure 4-25b are shown the six banana bonds that arise by considering all six occupied MOs simultaneously. To see how the four VAOs on carbon contribute to these bonds, consider Figure 4-26a. Here we hybridize an *spxpy* carbon VAO set with a carbon *pz* AO to give one $(spxpy)^1 (pz)^1$ set pointing three of its main lobes toward the right and the other set directing three such lobes toward the left. Each of these hybrid sets is 50% *spxpy* and 50% *pz* in character and each participates in three banana bonds with one oxygen, as shown in Figure 4-26b. Also shown in Figure 4-26b for both sets of *banana* bonds are the small antibonding contributions from the second oxygen which are required for orthogonality and

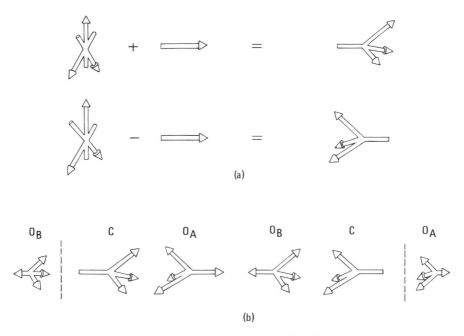

Figure 4-26. The hybrids $(spxpz)^1(pz)_r^1$ and $(spxpy)^1(pz)_l^1$ obtained by linear combination of $spxpy$ and $pz$ (a), and the six bonds in $CO_2$ (b), three of which are partially localized on the right side of the molecule and three on the left.

which reduce the bonding order of each of the banana bonds to $\frac{2}{3}$. Thus the overall bond order remains 2.0 per link as expected from the delocalized view with which we started.

From the preceding discussions it is clear that there is no one localized view which has two full bonding MOs in each link as well as two lp's on each oxygen. The electron dash structure in Figure 4-19a is therefore misleading. The point to be appreciated is that electron dash structures must be viewed with suspicion since they are frequently too simplistic to relate much information of any significance concerning the delocalized or the localized bonding.

## 4.4. Vibrational Modes for Linear Triatomics

To visualize the vibrational modes for a linear molecule YZY, we proceed in a manner analogous to that described in Chapter 3.3 for the vibration of $N_2$. In Figure 4-27 are the nine atomic vectors (AVs) on the Y and Z atoms which we must linearly combine into LCAVs permitted by GOs at the molecular center. As in the case of delocalized MOs, we can summarize the information on motions in a generator table. From Figure 4-27 we recognize that there are

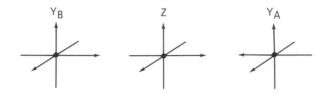

Figure 4-27. Atomic vector axis system for a linear YZY molecule.


Table 4-6. Generator Table for Linear YZY

| AV Equivalence Sets | GOs | | | | | |
|---|---|---|---|---|---|---|
| | $s$ | $pz$ | $px$ | $py$ | $dxz$ | $dyz$ |
| $Z(z)$ | | m | | | | |
| $Z(x)$, $Z(y)$ | | | m | m | | |
| $Y_A(z)$, $Y_B(z)$ | m | m | | | | |
| $Y_A(x)$, $Y_B(x)$, $Y_A(y)$, $Y_B(y)$ | | | m | m | m | m |

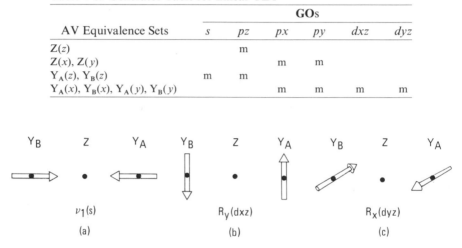

Figure 4-28. Symmetry and molecular motions generated by the (a) $s$, (b) $dxz$, and (c) $dyz$ **GO**s.

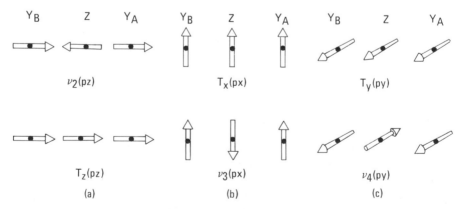

Figure 4-29. Symmetry and molecular motions obtained by linear combination of (a) $pz$-generated, (b) $px$-generated, and (c) $py$-generated symmetry motions.

the four AV equivalence sets listed in the left-hand column of Generator Table 4-6. From these sets are generated nine symmetry motions (SMs) denoted by the symbol "m". Here "m" stands for the "motion" having the same symmetry as the **GO** heading the column. The motions indicated in the first and last two columns of Table 4-6 are depicted in Figure 4-28. In Figure 4-28a a symmetric vibration ($v_1$) generated by $s$ is represented. The motions in Figure 4-28b and c are clearly rotations about the $y$ and $x$ axes of the center of mass (which also corresponds to atom Z), respectively. The $px$, $py$, and $pz$ **GO**s each generate two SMs and so linear combinations of these SMs must be taken. These are shown schematically in Figure 4-29. The three positive linear combinations of these SMs clearly lead to translations ($T$) in the directions $x$, $y$, and $z$ of the center of mass while the two negative combinations from the partner $px$ and $py$ SMs constitute degenerate bending vibrations of the molecule. We see that the expectation from the relationship $3n - 5 = 3(3) - 5 =$ four vibrational modes (two bending, two stretching) for a linear triatomic molecule is realized. Closely analogous modes arise in asymmetric linear triatomics XYZ. Although we generated the vibrations in Figures 4-28 and 4-29 in the order given by their subscripts, it should be noted that by convention in triatomic molecules, spectroscopists refer to the symmetric stretch (Figure 4-28a) as $v_1$, the bending motion (Figure 4-29b) as $v_2$, and the asymmetric stretch (Figure 4-29a) as $v_3$.

## Summary

By assuming linear geometries for $FHF^-$, FXeF, and $CO_2$, we saw from the delocalized views of these molecules that the sigma MOs of $FHF^-$ and FXeF (with no $d$ VAO participation on Xe) gave rise to a 3-center 4-electron bonding system. In this scheme two of the electrons are in a BMO and two are in an NBMO. In the corresponding localized views, these electron pairs could not be fully localized between two centers. By contrast, in $CO_2$ and FXeF (with $d$ VAO participation on Xe) both electron pairs in the sigma framework are in BMOs, and when these occupied MOs are localized, doubly occupied MOs spanning only two centers were realized. In formulating bonding views of FXeF and $CO_2$, we found that electron dash structures, while very useful, are sometimes misleading regarding the number of lone pairs to be preassigned.

Using generator orbitals we saw that the AV equivalence sets of a linear triatomic molecule give rise to SMs which can be summarized in a generator table. From the permitted linear combinations of these SMs, the vibrational modes were deduced.

In Table 4-7 is summarized an abbreviated form of our recipes for developing the delocalized view, a localized view, and the molecular motions. Discussions of all future examples will be based upon these stepwise procedures, which will be amplified in amended tables at the end of subsequent chapters.

Table 4-7. A Useful Summary of Procedures for Generating Pictorial Representations of Delocalized and Localized Bonding Views of a Molecule and for Visualizing its Vibrational Modes

| Delocalized View | Localized View | Normal Modes |
|---|---|---|
| 1. Decide on best electron dash structure. | 1. Decide on which set(s) of doubly occupied delocalized MOs to localize. | 1. Identify the motional atomic vector (AV) equivalence sets and list them in the generator table. |
| 2. Establish the molecular geometry. | | |
| | 2. Hybridize the **GO**s which give rise to the sets(s) in 1. | |
| 3. After preassigning $lps$ to ($s$) and suitably oriented ($p$) VAOs, identify all the VAO equivalence sets and list them in the generator table. | | 2. Using **GO**s, sketch the symmetry motions (SMs). |
| | 3. Using the hybridized **GO**s, sketch the SOs. | 3. Fill in the rest of the generator table. (i.e., **GO**s and m entries). |
| | 4. Sketch the MOs, being aware of the consequences of a central atom on which the full set of SOs (hybridized (VAOs) is not generated by the hybridized **GO** set. | |
| 4. Using **GO**s, sketch all the SOs, recalling that #SOs = #VAOs. | | |
| 5. Fill in rest of generator table (i.e., the necessary **GO**s and the $n$, $a$, or $b$ entries). | | 4. Take linear combinations of SMs generated by the same **GO** and sketch all the molecular motions. |
| | | 5. Identify the normal vibrational modes. |
| 6. Take linear combinations of SOs generated by the same **GO** and sketch all the MOs. | 5. Draw the resulting MO energy level diagram and occupy the appropriate levels with electrons. | |
| 7. Draw an MO energy level diagram and occupy the appropriate levels with electrons. | 6. Verify that the bond order per link is the same as in the delocalized view. | |
| 8. Deduce the bond order per link. | | |

## PROBLEMS

1. For the linear species below, draw the electron dash structure and generate delocalized and localized views of the bonding. (That is, sketch the LCAOs, give the **GO** tables, draw properly labeled MO energy level diagrams, and comment on the bonding.)

   (a) $BeH_2$; (b) $CH_2$ (paramagnetic); (c) $I_3^-$; (d) $CS_2$

2. Repeat Problem 1c including ($d$) VAOs for $I_3^-$.

3. Sketch the MO energy level diagrams corresponding to the localized views of $CO_2$ depicted in Figures 4-24b and 4-25b.

4. Generate delocalized and localized views of linear OCS.

5. Using the **GO** approach, deduce the vibrational modes of HCN.

6. Using the **GO** approach, generate a delocalized view of $(CN)_2$ and also deduce its vibrational modes. [Hint: Treat the system as two symmetrical parts, namely, two carbons and two nitrogens, and take linear combinations of symmetry orbitals (or atomic vectors) which are generated by the same **GO**.]

7. Give the delocalized view of $N_3^+$ in the linear geometry wherein it possesses two unpaired electrons.

# CHAPTER 5

# Triangular and Related Molecules

In this chapter we will examine $H_3^+$, $N_3^+$, $CH_3^+$, and $BF_3$. The first two species are truly triangular in that they have an atom at each of the three corners of an equilateral triangle. Although the latter two molecules also have three identical atoms on the corners of a triangle, they possess a central atom as well.

## 5.1. $H_3^+$

This perhaps strange-looking ion was discovered by Thompson in hydrogen discharge studies in 1912. Other triangular species which have been studied in the mass spectrometer are $Li_3^+$, $Na_3^+$, $Na_2Li$, $NaLi_2^+$, $Li_2H^+$, $Na_2H^+$, and $NaLiH^+$. Theoretical calculations show that the equilibrium $H_3^+$ configuration is an equilateral triangle with interatomic distances comparable to that in $H_2$.

To begin our discussion of the delocalized bonding in $H_3^+$, let us adopt the axis system in Figure 5–1. The dots denote atom centers and the triangle denotes our **GO** center. From the symmetry of $H_3^+$ we see that all of the $(H1s)$ VAOs belong to the same equivalence set in Table 5-1. For example, rotation by 120° about the $Z$ axis brings $(H_A1s)$ into $(H_B1s)$, and $(H_B1s)$ into $(H_C1s)$. From Figure 5-2 it is seen that the three SOs expected from the three VAOs are generated by $s$, $px$, and $py$ GOs. Note in Figure 5-2b that $H_A$ is unable to contribute to $\tilde{\sigma}(py)$ since the $py$ **GO** and $(H_A1s)$ behave differently with respect to reflection in the $xz$ plane, one of the symmetry operations in equilateral $H_3^+$. Notice also in Figure 5-2c that the $(H_B1s)$ and $(H_C1s)$ contributions are only half as large as that of $(H_A1s)$. A way to visualize the reason for this result is to consider Figure 5-3. The $(1s)$ VAO of $H_A$ is seen to be called

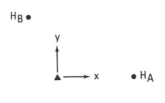

Figure 5-1. Axis system for $H_3^+$.

Table 5-1. Generator Table for $H_3^+$

|  | GOs | | |
| --- | --- | --- | --- |
| VAO Equivalence Sets | $s$ | $px$ | $py$ |
| $Hs = (H_A 1s), (H_B 1s), (H_C 1s)$ | b | a | a |

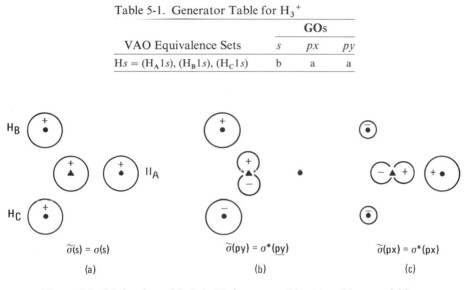

Figure 5-2. Molecular orbitals in $H_3^+$ generated by (a) $s$, (b) $py$, and (c) $px$.

in as fully as possible by $px$ and so we give this VAO a coefficient of 1. If we place an $(H1s)$ VAO at point $P$ directly above $px$, then this **GO** is unable to call in a contribution from such a VAO because this VAO does not have a node where the **GO** does. Thus the coefficient of 1 for $(H_A 1s)$ must decrease to zero if a $(1s)$ orbital should move from the $H_A$ position to point $P$. By the same token, this coefficient must rise negatively to $-1$ if the VAO at $P$ moves to $R$. The path of a $(1s)$ VAO around the circle back to $H_A$ again is seen to reflect a sine function where the coefficient for the $(1s)$ orbital at $H_B$ is given by sin $120° = -1/2$. Similarly, the coefficient of $(H_C 1s)$ is $-1/2$. Another way to rationalize the coefficients implied in Figure 5-2c is to recognize that they are required for orthogonality of this MO with the other two.

Since there is only one SO generated by each **GO**, the SOs are MOs. Thus,

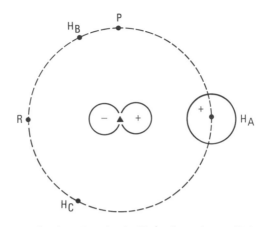

Figure 5-3. Diagram of points $P$ and $R$ in $H_3^+$ where the coefficients of (H1$s$) VAOs are 0 and $-1$ for an SO generated by $px$.

because there is no central atom in this molecule, the atoms on the corners of the triangle must bind to one another. The bonding and antibonding nature of these MOs indicated in Table 5-1 can be pictorially recognized from Figure 5-2. The MO $\sigma(s)$ in Figure 5-2a is manifestly bonding between all pairs of atoms. The antibonding interaction between $(H_B 1s)$ and $(H_C 1s)$ in Figure 5-2b clearly marks this MO as antibonding. The antibonding interactions between $(H_A 1s)$ and its two neighbors in $\sigma^*(px)$ (Figure 5-2c) exceed the single bonding interaction between $(H_B 1s)$ and $(H_C 1s)$, thus justifying the labeling of this MO as $\sigma^*(px)$.

Before we can proceed to draw an MO energy level diagram, however, we need to know whether the energies of $\sigma^*(py)$ and $\sigma^*(px)$ are the same or different. The orbitals certainly *look* different. What possible reason could there be for degeneracy? To appreciate the fact that these two ABMOs indeed are degenerate, we need to reconsider the triangular symmetry of the molecule. A triangle has a threefold axis of symmetry, which means that any point within the molecule when rotated by 120°, 240°, or 360° feels the same potential. The same is true if we reflect a point through the $xy$ plane, for example. Now the $\sigma(s)$ MO and, of course, the $s$ **GO** which generates it have this full symmetry of the ion, too. This is not true of the ABMOs *individually*, however, nor of the $px$ and $py$ **GO**s which generate them. But it is easy to see that rotation of a $px$ **GO** by 120°, for example, creates a new $p$ orbital which must be a linear combination of $px$ and $py$. In other words the $px$ and $py$ **GO**s *taken together* do contain the full symmetry of our triangular molecule and this is also true for the two SOs they generate. This is easily visualized by squaring the wavefunctions of $(px)$ and $(py)$ and summing their electron densities. A three-dimensional plot of the sum resembles a perfectly circular donut lying in the xy plane. Such a shape has the full symmetry of a triangle or, for that matter, any regular polygon (Chapter 7). Because the $px$- and $py$-

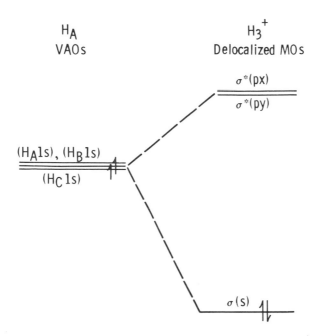

Figure 5-4. Delocalized MO energy level diagram for $H_3{}^+$.

generated SOs are generated by **GO**s which are equal partners in containing the full symmetry of the $H_3{}^+$ ion, they also have equal energies. It is always true that when a molecule has an SO which does not contain the full symmetry of the molecule, there is at least one other partner SO of equal energy which also does not contain the full symmetry of the molecule. *Moreover, partner SOs are always generated from one VAO set by GOs having the same l quantum number.* In atoms, any ($s$) orbital contains the full symmetry of the atom. A ($p$) orbital by itself obviously does not but when all three ($p$) orbitals are taken together, their combined electron density generates a perfectly spherical charge density. Thus the full set of AOs with a given l quantum number always contains the symmetry of a sphere. An $s$-generated SO never has a partner since a sphere (the highest symmetry there is) contains the full symmetry of the atom. A $p$-generated SO can have one or two partners, a $d$-generated SO can have up to four partners, etc. It turns out, however, that partner sets containing more than three members are rare.

The delocalized MO energy level diagram of $H_3{}^+$ shown in Figure 5-4 indicates that the bond order per link is 1/3. Since there is only one occupied MO in this molecule, no further localization is possible. The bond is a 3-center 2-electron bond, thus making it impossible to draw an electron dash structure. However, we can draw the three so-called *resonance* structures shown in Figure 5-5a–c. Resonance structures are electron dash structures which can be drawn by moving electrons around the structure while leaving all atoms

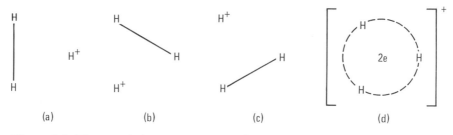

Figure 5-5. Electron dash structures for $H_3^+$ [(a), (b), (c)] and an average of these structures (d).

stationary. Individually the resonance structures of $H_3^+$ are unsatisfactory, but taken together as an average structure (which is depicted in Figure 5-5d) they confirm the experimental fact that all three atoms are chemically equivalent.

## 5.2. $N_3^+$

This ion formed by electron impact of $N_2$ at low pressures can also be studied in the mass spectrometer. Theoretical calculations reveal it to have all its electrons paired in the equilateral triangular geometry. Although a linear geometry (with two unpaired electrons) is calculated to be of lower energy for this ion, we will examine the metastable triangular arrangement since the bonding pattern in this geometry has important implications for more common molecules such as $BF_3$. The $N_3^+$ ion is isoelectronic with the cyclopropenium ion $(CH)_3^+$, which is derived from cyclopropene by abstraction of a hydride ion.

    As with $H_3^+$, attempts to draw a single electron dash structure of $N_3^+$ and $(CH)_3^+$ are less than pleasing and we must draw resonance structures (e.g., Figure 5-6a, b, for each of which there are two additional similar structures achieved by moving the double bond and the charge around the rings). It is interesting to note that $N_3^+$ and $(CH)_3^+$ are isoelectronic (i.e., they have the same number of valence electrons). Moreover, $(CH)_3^+$ can be thought of as resulting from the movement of a proton from each nitrogen nucleus radially outward to the hydrogen position where it converts a nitrogen lp to a bonding electron pair (bp).

    In developing the delocalized view of $N_3^+$, it is reasonable first to assign the lone pair on each nitrogen to an (N2s) VAO. Using the axis system in Figure 5-7 we construct the equivalence sets in Generator Table 5-2. We note, however, that the N$pr$ and N$pt$ equivalence sets in this table denote radially and tangentially directed ($p$) VAOs, respectively. Generation of the SOs in the N$pr$ equivalence set is done exactly in the same manner as for $H_3^+$. As shown in Figure 5-8, the $pz$, $dxz$, and $dyz$ GOs generate three SOs among the VAOs of

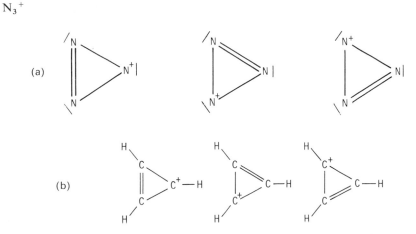

Figure 5-6. Electron dash resonance structures for $N_3^+$ (a) and $(CH)_3^+$ (b).

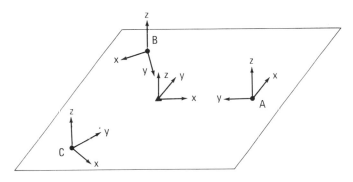

Figure 5-7. Axis system for $N_3^+$.

Table 5-2  Generator Table for $N_3^+$

| VAO Equivalence Sets | GOs | | | | | | |
|---|---|---|---|---|---|---|---|
| | $s$ | $px$ | $py$ | $pz$ | $dxz$ | $dyz$ | $f3''$ |
| $Npr = (N_A2py), (N_B2py), (N_C2py)$ | b | a | a | | | | |
| $Npt = (N_A2px), (N_B2px), (N_C2px)$ | | b | b | | | | a |
| $Np\pi = (N_A2pz), (N_B2pz), (N_C2pz)$ | | | | b | a | a | |

the $Np\pi$ set. Since these are the only SOs generated by these three GOs, these SOs are MOs. Using the same reasoning as for the MOs of $H_3^+$, the bonding and antibonding natures of the MOs in Figure 5-8, the coefficients (sizes) of the VAOs, and the degeneracy of the $\pi^*$ *partner-set* MOs should be clear to the reader. Notice that in $\pi^*(dxz)$ the size of $(N_A2pz)$ was doubled instead of reducing the sizes of $(N_B2pz)$ and $(N_C2pz)$ in half. This produces no difference in the overall results because the wave functions can be normalized later.

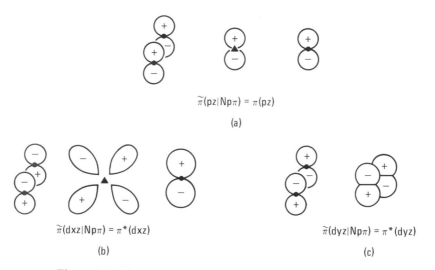

$$\tilde{\pi}(pz|Np\pi) = \pi(pz)$$

(a)

$$\tilde{\pi}(dxz|Np\pi) = \pi^*(dxz) \qquad\qquad\qquad\qquad \tilde{\pi}(dyz|Np\pi) = \pi^*(dyz)$$

(b)                                                                                      (c)

Figure 5-8. The $\pi$ MOs generated in the $Np\pi$ equivalence set of $N_3{}^+$.

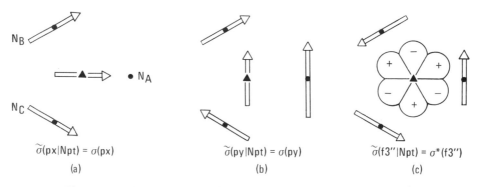

$$\tilde{\sigma}(px|Npt) = \sigma(px) \qquad\qquad \tilde{\sigma}(py|Npt) = \sigma(py) \qquad\qquad \tilde{\sigma}(f3''|Npt) = \sigma^*(f3'')$$

(a)                                              (b)                                              (c)

Figure 5-9. The $\sigma$ MOs generated in the $Npt$ equivalence set of $N_3{}^+$.

From Figure 5-9 it is seen that $px$ and $py$ generate a partner set of MOs which, in contrast to the analogous situation in $H_3{}^+$, now gives a pair of bonding (rather than antibonding) MOs. That is, $\sigma(px)$ is manifestly bonding between the lobes directed toward one another between atoms B and C. While this is not true for $\sigma(py)$, the two bonding interactions between atoms A and B and A and C clearly outweigh the single antibonding interaction between B and C. The remaining SO in the $Npt$ equivalence set is found to be generated by $f3''$. No **GO** simpler than $f3''$ gives this new SO. That $f3''$ is indeed correct here is seen by recognizing that since only one SO remains to be generated, it cannot be degenerate with another (i.e., have a partner) and indeed $f3''$ is seen from Figure 5-9c to contain the full symmetry of a triangle. The MO $\sigma(f3'')$ is seen to be antibonding, rather than bonding as is the case with the *non-*

degenerate MO in the N$pr$ and N$p\pi$ equivalence sets. Except for the $px$- and $py$-generated MOs in Table 5-2, all the MOs are generated by a single **GO**. By allowing the $px$- and $py$-generated antibonding and bonding MO pairs to interact, further enhancement of their antibonding and bonding character will occur.

There is a point concerning our choice of the axis system for $N_3^+$ which needs further comment. If we had not chosen to point the $x$ axis toward one of the nitrogen atoms, the $f3''$ GO could not have been utilized. Thus it would have been necessary to hybridize $f3'$ with $f3''$ to obtain a linear combination which would have a planar node intersecting each N atom. In future examples involving polygonal arrays of atoms in the $xy$ plane, we will therefore generally point the $x$ axis toward one of the atoms. The square array is an exception which we will treat in Chapter 7.

In $N_3^+$ we have fourteen valence electrons, six of which we have preassigned to the three (N2$s$) VAOs. The remaining eight are expected to occupy the four BMOs we have generated. In Figure 5-10 a delocalized MO energy level diagram is given. In this diagram all four BMOs are 3-center 2-electron bonds.

Before we go on to the localized view, let us modify our delocalized view somewhat in order to make our method for the treatment of future molecular

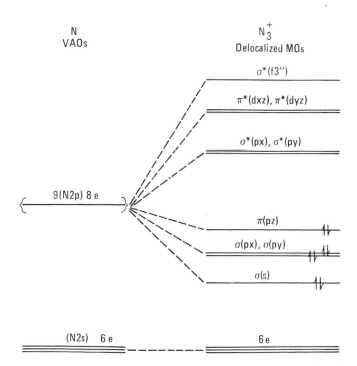

Figure 5-10. Delocalized MO energy level diagram for $N_3^+$.

Table 5-3. Generator Table for $N_3^+$ (without lp Preassignment)

| VAO Equivalence Sets | GOs | | | | | | |
|---|---|---|---|---|---|---|---|
| | $s$ | $px$ | $py$ | $pz$ | $dxz$ | $dyz$ | $f3''$ |
| $Ns = (N_A2s), (N_B2s), (N_C2s)$ | b | a | a | | | | |
| $Npr = (N_A2py), (N_B2py), (N_C2py)$ | b | a | a | | | | |
| $Npt = (N_A2px), (N_B2px), (N_C2px)$ | | b | b | | | | a |
| $Np\pi = (N_A2pz), (N_B2pz), (N_C2pz)$ | | | | b | a | a | |

Table 5-4. Generator Table for $N_3^+$ (without Preassignmentof lp's in $Nspo$ Hybrids)

| VAO Equivalence Sets | GOs | | | | | | |
|---|---|---|---|---|---|---|---|
| | $s$ | $px$ | $py$ | $pz$ | $dxz$ | $dyz$ | $f3''$ |
| $Nspi = (N_Aspi), (N_Bspi), (N_Cspi)$ | b | a | a | | | | |
| $Npt = (N_A2px), (N_B2px), (N_C2px)$ | | b | b | | | | a |
| $Np\pi = (N_A2pz), (N_B2pz), (N_C2pz)$ | | | | b | a | a | |

ring systems more convenient to apply. If we had not preassigned our lone pair orbitals to (N2s) VAOs, we would have had an $Ns$ equivalence set from which three SOs (MOs) would have been generated by $s$, $px$, and $py$, as shown in Generator Table 5-3. Under these circumstances, the linear combinations of the MOs generated by $px$, for example, would lead to three MOs in which $\sigma_s(px)$ would be essentially nonbonding compared to the other two [since the (N2s) VAO lies so low in energy], $\sigma_{px}(px)$ would still be quite bonding, and $\sigma_{py}^*(px)$ would be quite antibonding. We could, of course, also find suitable orthogonal linear combinations of these MOs in which the $s$-dominated $\sigma_s^0(px)$ MO is entirely free of admixtures from $Npt$ and thus contains $Ns$ and $Npr$ contributions only. That would mean that the predominant $Npt$ contribution to $\sigma_{px}(px)$ would become even stronger. This hybridization of the MOs generated by $px$ (and similarly by $py$) is convenient since it creates a set of digonal ($s$)-dominated $Nsp$ hybrids pointed radially outward ($Nspo$), and a set of digonal ($p$)-dominated $Nsp$ hybrids pointed radially inward ($Nspi$). The $Npt$ set is thereby preserved intact. We could, of course, have arrived essentially at the same point by hybridizing the nitrogen atoms at the start so that each one has an ($s$)-dominated ($Nspo$) and an ($Nspi$) VAO *and preassigning a lone pair to each* ($Nspo$). This procedure was useful in $FHF^-$, for example, and it will find great use in future applications. The modified generator table given as Table 5-4 and the MO energy level diagram in Figure 5-11 reflect this modification. We could, of course, include the $Nspo$ equivalence set in Table 5-4 in which case the $s$ and the $px$, and $py$ **GO**s would generate a bonding and two antibonding SOs, respectively. However, since all three of these SOs (MOs) are occupied, we can always recast these MOs into the ($Nspo$) VAOs (with small antibonding contributions from the other two atoms). For convenience we therefore chose to keep lone pairs localized in VAOs and to a first

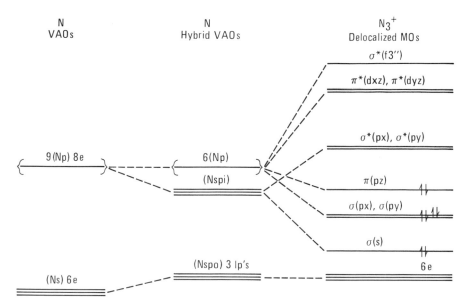

Figure 5-11. Delocalized energy level diagram for $N_3^+$ with preassignment of lp's in N$spo$ hybrids.

approximation this is valid. This procedure is reflected in Figure 5-11. In the MO energy level diagram, the lp's are seen to have mainly ($s$) character in the (N$spo$) VAOs and the ($p$)-dominated $\sigma^*(px)$ and $\sigma^*(py)$ MOs have dropped in energy with the admixture of the ($s$) character, as has the $\sigma(s)$.

To obtain the localized picture of $N_3^+$, we begin by hybridizing the $s, px, py$ **GO** set which generated the $\sigma$ BMOs in Figure 5-11. This $sp^2$ hybrid set points its main lobes toward the three atom centers. Since we always prefer that the localized MOs be concentrated *between* atom centers, we take advantage of the concept developed in Chapter 2 that we can recast our present $sp^2$ **GO** set into one rotated by 60°. The main lobes now are directed at the bond mid-points. The effect of each lobe of this hybrid **GO** set is to call in an (N$sp^\pi$) hybrid from each nitrogen (Figure 5-12a) made up of an (N$spi$) and an (N$pt$) VAO (Figure 5-12b). The three localized MOs composed mainly of contributions from an (N$sp^\pi$) VAO on each pair of atoms are depicted in Figure 5-12c. It is clear that the (N$sp^\pi$) hybrids are not 180° apart (digonal, $\alpha = 1$) since an entire (N$pt$) VAO is involved plus the dominating ($p$) contribution in (N$spi$). Moreover, only part of the (N$s$) VAO contributes to the (N$sp^\pi$) VAO since part of the (N$s$) orbital set is already contributing to the lp (N$spo$) orbital. Further-more, the (N$sp^\pi$) orbitals cannot be facing directly toward each other along the internuclear axis since that would require an angle of 60° between this pair of these hybrids on each atom. Even with pure $p$ character, two orthogonal ($p$) VAOs must be 90° apart. Since there is only one occupied $\pi$ delocalized MO, no further localization is possible. The only difference between the

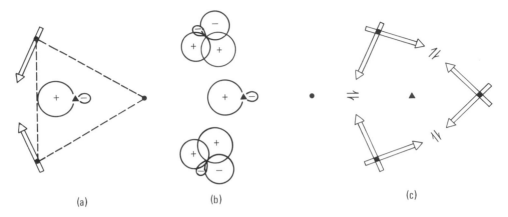

(a)                              (b)                                (c)

Figure 5-12. Generation by an $sp^2$ **GO** of a localized MO in $N_3^+$ between two nitrogens using a combination of their $Nspi$ and $Npt$ VAOs (a), which is shown in more detail in (b). In (c) is shown a pictorial representation of all three localized MOs.

delocalized MO energy level diagram in Figure 5-11 and its localized counterpart is that the occupied $\sigma$ delocalized MOs are recast into three degenerate localized BMOs $\hat{\sigma}_A$, $\hat{\sigma}_B$, $\hat{\sigma}_C$ which lie two-thirds of the way up from $\sigma(s)$ to the degenerate $\sigma(px)$, $\sigma(py)$ pair.

## 5.3. Electron Dash Structures and Molecular Geometry

Before discussing the bonding in more complex molecules, it is instructive to make some further observations concerning electron dash structures. For if we can draw a reasonable electron dash structure, the geometrical structure can generally be inferred from it and the formulation of the bonding patterns is greatly facilitated. How can we draw an electron dash structure, given a formula? Taking $XeF_2$ as an example, we could write the two electron dash structures in Figure 5-13a and b if we did not happen to know the experimentally determined structure. A simple calculation of the so-called *formal charges* shown above each atom permits us to make the correct choice. *The formal charge is given by the difference between the number of valence electrons in the neutral atom and the sum of the lp electrons and half the number of bonding pair electrons of the atom in the electron dash structure.* In general, the electron dash structure with the lowest set of formal charges is the one to choose (i.e., Figure 5-13a). If there is no alternative to having more than one formal charge of the same sign, then separating these charges by the largest number of atoms possible affords the best electron dash structure. How does the dash structure in Figure 5-13a allow us to deduce the geometry of the molecule? Electron dash structures tend to reflect the localized MO picture. We have seen that in such a bonding view the localized valence MOs tend to avoid each other

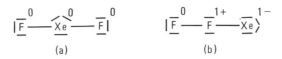

Figure 5-13. Electron dash structures for two forms of XeF$_2$.

in space and that lp localized MOs contain more $s$ character and hence command a greater spherical angle around a nucleus than do localized BMOs. The recognition by Gillespie and Nyholm that actual bond angles adjust themselves to this behavior led to the "valence state electron pair repulsion" (VSEPR) rules which state that the order of "repulsion" effects is lp–lp > lp–bp > bp–bp. Actually, electrostatic repulsions probably play a minor role. The mutual avoidance of localized MOs is largely governed by antibonding effects arising from the Pauli principle (which forces different pairs into different orbitals) and antibonding overlap effects between doubly filled orbitals.

In triatomic molecules, for example, it is possible to have zero (HBeH), one (ONO$^-$), two (H$_2$O), and three (FXeF) lp's on the central atom. The VSEPR rules easily predict a 180° bond angle for HBeH, < 120° for ONO$^-$, and < 109°28′ for H$_2$O. For five pairs of electrons around xenon in FXeF (three lp's and two bp's if ($d$) VAO participation on Xe is assumed), calculations show that if all "repelled" one another equally, they would arrange themselves on the surface of a sphere at the apices of a trigonal bipyramid or at the corners of a square pyramid, which is generally not much higher in energy than a trigonal bipyramid. In most examples, however, the trigonal bipyramidal geometry is adopted. Since lp's "repel" other lp's most strongly, they arrange themselves 120° apart from one another around the equatorial plane in the trigonal bipyramid while the bp's take up the remaining two positions on the axis, thus placing themselves 90° apart from the lp's. In H$_3$$^+$ the VSEPR rules apply in a negative sense; that is, there are no electron pair repulsions to prevent ring closure. Closing the ring is favorable here because it leads to three delocalized bonds instead of only two in a linear or bent arrangement.

## 5.4. BF$_3$

Examples of molecules with a central atom bonded to a triangular array of other atoms are BX$_3$, NO$_3$$^-$, CO$_3$$^{2-}$, BO$_3$$^{3-}$, CH$_3$, NH$_3$$^+$, BeH$_3$$^-$, CH$_3$$^+$, BH$_3$, BH$_3$$^-$, InX$_3$, and GaX$_3$. We will choose to examine BF$_3$ since it introduces another useful feature of electron dash structures. In Figure 5-14a is an electron dash structure which, although it does leave zero formal charges on the atoms, does not obey the Lewis octet rule. At a more fundamental level than we usually learn this rule, it says that when four VAOs are available on an atom, they *all* become involved in the bonding scheme. Placing a double bond in one of the links (Figure 5-14b) by using one of the fluorine lp's shown in Figure 5-14a allows us to write in an analogous way two more equivalent structures with the double bonds in the other two links. The three resonance

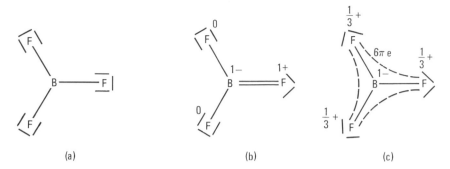

(a)                                    (b)                                    (c)

Figure 5-14. Electron dash structures for $BF_3$ in which the octet rule is only partially obeyed (a) and completely obeyed (b) but with the generation of formal charges. In (c) is an average electron dash structure for the three resonance structures implied by the one in (b).

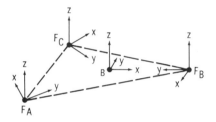

Figure 5-15. Axis system for $BF_3$.

structures which result involve a formal charge of $-1$ on the boron and a $+1/3$ charge on each fluorine. The presence of the double bond in these resonance structures implies that this bond (which is in a $\pi$ MO perpendicular to the molecular plane) is delocalized over all four atoms. The term resonance is taken to mean that if a single electron dash structure were to be drawn, it would be an average of the three resonance forms implied by Figure 5-14b. Such an average structure (Figure 5-14c) would have to involve a $(2pz)$ VAO on each fluorine and the $(B2pz)$ VAO, as indicated by the axis system in Figure 5-15. Since each of the $(F2pz)$ VAOs is occupied by lp's in Figure 5-14a, there must be a total of six electrons in the delocalized $\pi$ system since the $(F2pz)$ VAOs are now involved in the delocalized $\pi$ MOs. Because of the presence of the formal charges in Figure 5-14b and c, there is a polarization of the bonds in each of the links toward the fluorines. This is not unexpected since fluorine is more electronegative than boron. The presence of bond polarization means that we cannot discount the contribution of the three ionic resonance structures implied by Figure 5-16.

It is clear that the molecule must be planar since the $\sigma$ bond pairs "repel" one another to 120°. Pi electron density also "repels" and its even distribution among all three links is expected to reinforce the trigonal planar geometry of

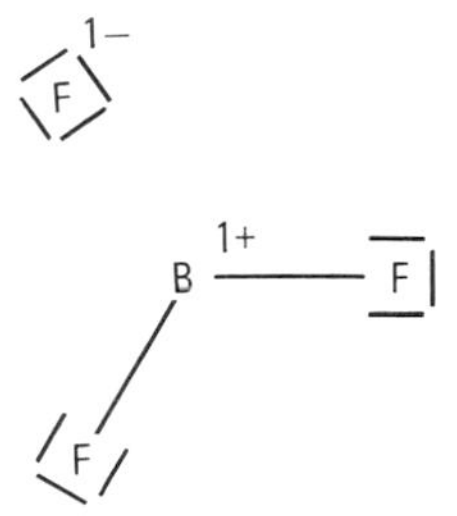

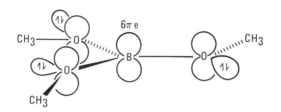

Figure 5-16. One of three ionic resonance structures for BF$_3$.

Figure 5-17. Structure of B(OCH$_3$)$_3$.

the molecule. In Chapter 1 it was pointed out that double bonds are shorter than single bonds. Indeed the B—F bond in BF$_3$ is found experimentally to be shorter than expected for a B—F single bond. Could such bond shortening arise solely from coulombic attraction in the ionic resonance structures (Figure 5-16)? The answer is probably not, since $\pi$ bonds are expected to experience restricted bond rotation while ionic interactions are not. Although restricted bond rotation is not possible to detect experimentally in BF$_3$, it is implied by the planar arrangement found in the BO$_3$C$_3$ portion of B(OCH$_3$)$_3$ (Figure 5-17), which is isoelectronic with BF$_3$ except that one of the lone pairs on the fluorines is converted to an O—CH$_3$ bond. If delocalized $\pi$ bonding were not important in this molecule, a nonplanar conformation unfavorable to delocalization might be expected.

In developing the delocalized view of BF$_3$ we begin by assigning one lp on each fluorine to an (F2$s$) VAO and a second fluorine lp to an (F2$px$) VAO. In this way, the (F2$pz$) and (F2$py$) VAOs are reserved for the perpendicular $\pi$ and the radial $\sigma$ systems, respectively, as shown in Generator Table 5-5. For convenience we will henceforth also indicate in the generator tables the number of VAO electrons to be accounted for over and above the preassigned lp's. Since the $s$-, $pz$-, $dxz$-, and $dyz$-generated MOs are analogous to those generated for N$_3{}^+$, their pictorial representations will not be repeated here. The new features in BF$_3$ are the pairs of SOs generated by $s$, $px$, $py$, and $pz$ which we must linearly combine, and this is shown in Figure 5-18. In addition, there are the $\pi^0(dxz)$ and $\pi^0(dyz)$ NBMOs. The average bond order from the average electron dash structure is expected to be 4/3 and this is confirmed by the MO energy level diagram in Figure 5-19.

Table 5-5. Generator Table for BF$_3$ (12e$^-$)

| VAO Equivalence Sets | GOs | | | | | |
|---|---|---|---|---|---|---|
| | $s$ | $px$ | $py$ | $pz$ | $dxz$ | $dyz$ |
| B$s$ = (B2$s$) | n | | | | | |
| (B$p\sigma$ = (B2$px$), (B2$py$) | | n | n | | | |
| B$p\pi$ = (B2$pz$) | | | | n | | |
| F$pr$ = (F$_A$2$py$), (F$_B$2$py$), (F$_C$2$py$) | n | n | n | | | |
| F$p\pi$ = (F$_A$2$pz$), (F$_B$2$pz$), (F$_C$2$pz$) | | | | n | n | n |

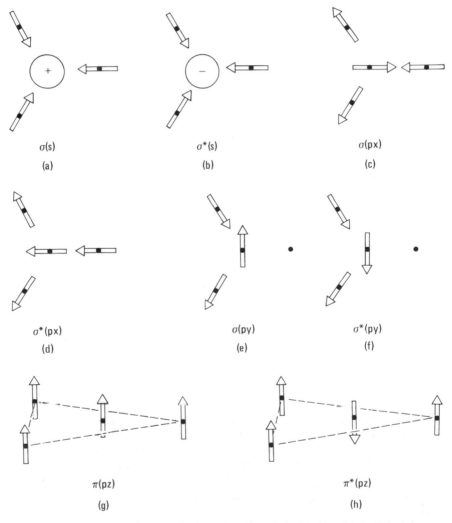

Figure 5-18. MOs generated by **GO**s in the plane [(a), (b), (c), (d), (e), (f)] and perpendicular to the plane [(g), (h)] of BF$_3$.

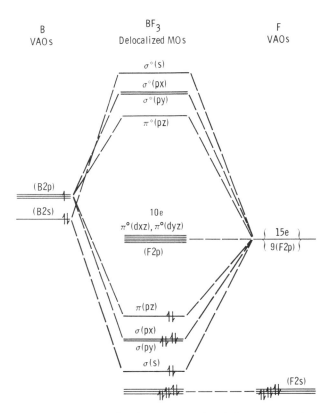

Figure 5-19. Delocalized MO energy level diagram for BF₃.

For greater ease in visualizing the localized view, we choose to localize the σ and π MO sets separately. From our usual procedure we conclude that the main lobes of an $sp^2$ **GO** set are directed toward the fluorines and they call in a set of localized MOs $\hat{\sigma}_A$, $\hat{\sigma}_B$, $\hat{\sigma}_C$, each composed mainly of an ($sp^2$) boron VAO and a fluorine ($2py$) VAO. The three π MOs in Figure 5-19 are similarly localized except that the $pzdxzdyz$ **GO** set (of which one is depicted in Figure 5-20a) generates only the ($2pz$) VAO on boron (Figure 5-20b). As we have come to expect from such a situation, the localized bond is partially cancelled by antibonding contributions from the other two atoms in Figure 5-20b in order to preserve orthogonality. Each localized π BMO contributes a bond order of $+1$ in one B—F link and a bond order of $-1/3$ in each of the two others, resulting in an overall π bond order of 1/3 in each link, as was found in the delocalized view. In Figure 5-21 is shown the localized MO energy level diagram. In it the fluorine lp's are depicted as hybrid VAOs. Such hybrids are, of course, more localized than $s$ and $p$ lp's in canonical VAOs.

In closing this chapter we briefly consider the molecular motions available to a triangular molecule with and without a central atom. For the latter case these motions are summarized in Figure 5-22 and Generator Table 5-6. In this

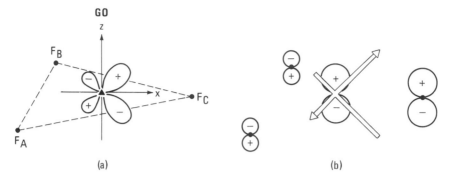

Figure 5-20. The *pzdxzdyz* **GO** in BF$_3$(a) which generates a partialy localized $\pi$ BMO (b) from the (B2*pz*) and the (F2*pz*) VAOs. Note the "double headed" and "double tailed" notation for the *pd²* **GO**.

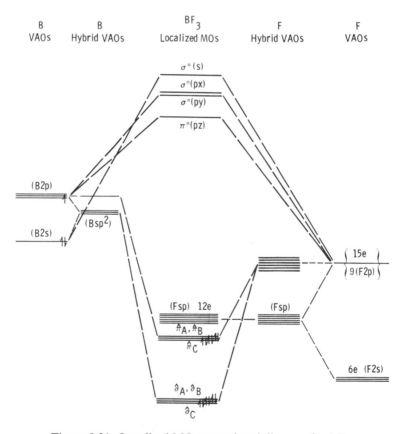

Figure 5-21. Localized MO energy bond diagram for BF$_3$.

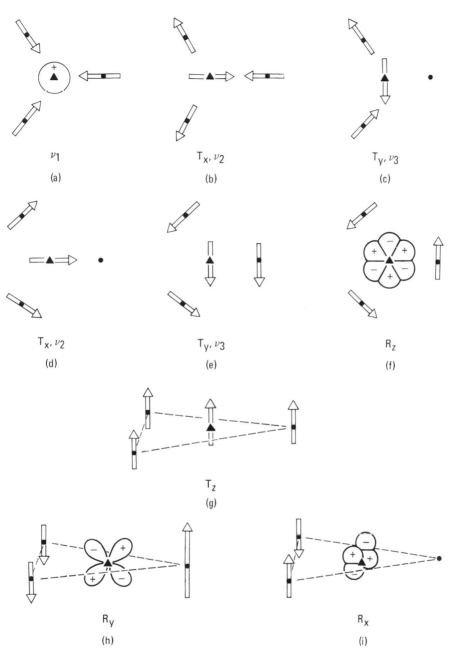

Figure 5-22. Molecular motions for a triangular Z$_3$ molecule.

Table 5-6. Generator Table for Triangular $Z_3$

| AV Equivalence Sets | GOs | | | | | | |
|---|---|---|---|---|---|---|---|
| | $s$ | $px$ | $py$ | $pz$ | $dxz$ | $dyz$ | $f3''$ |
| $Zy$ | m | m | m | | | | |
| $Zx$ | | m | m | | | | m |
| $Zz$ | | | | m | m | m | |

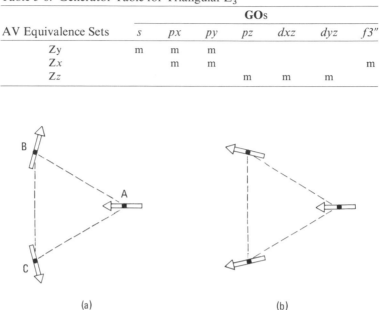

(a)                                                  (b)

Figure 5-23. Linear combinations of the atomic vectors of Figure 5-22b and d. In (a) is shown the sum of the vectors and in (b) is shown the difference.

and future generator tables we will frequently list only the generic symbols for the equivalence sets and not their individual members. In Figure 5-22a–c and d–f are shown the top views of GO-induced SMs among the radial and tangential AVs, respectively. In Figure 5-22g–i are shown perspective views of the SMs among the "$\pi$" AVs. It is clear from Figure 5-22 that the $s$ GO leads to a vibration, that $pz$ generates a translation, and that $dxz$, $dyz$, and $f3''$ each provide a rotation. Note that the larger AV contribution on atom A (by a factor of 2) generated by $dxz$ compared to the two smaller ones on B and C in Figure 5-22h is in agreement with the fact that the momentum of the first atom must be twice that of the other two in order for the angular momentum in the positive $z$ direction to be equal to the sum of the angular momenta of the two in the negative $z$ direction. Since $px$ and $py$ each generate two SMs, linear combinations must be taken. In Figure 5-23a is shown the superposition of the two AVs of Figure 5-22b and d as a sum and in Figure 5-23b it is shown as a difference. By taking the linear combination in Figure 5-23b with a somewhat higher (negative) contribution of the SM in Figure 5-22d, the vectorial sums of the motions on atoms B and C would be parallel with the movement of atom A (Figure 5-24) and we would have a translation along the $-x$ direction of the center of mass. The other linear combination (Figure 5-23a) represents a vibration that stretches the $N_B–N_C$ bond while the $N_B–N_A$ and $N_C–N_A$ bonds are compressed. In a similar manner the linear

Figure 5-24. Linear combination shown in Figure 5-23b with a more negative contribution of the SM in Figure 5-22d.

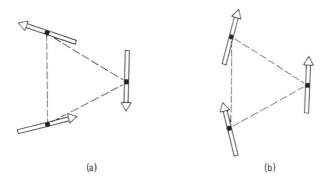

(a)                                              (b)

Figure 5-25. Linear combinations of the atomic vectors in Figure 5-22c and e. In (a) is shown the sum and in (b) is shown the difference.

Table 5-7.  Generator Table for Trigonal Planar YZ$_3$

| AV Equivalence Sets | GOs | | | | | | |
|---|---|---|---|---|---|---|---|
| | $s$ | $px$ | $py$ | $pz$ | $dxz$ | $dyz$ | $f3''$ |
| Y$xy$ | | m | m | | | | |
| Y$z$ | | | | m | | | |
| Z$y$ | m | m | m | | | | |
| Z$x$ | | m | m | | | | m |
| Z$z$ | | | | m | m | m | |

combinations of the SMs in Figure 5-22c and e shown in Figure 5-25a and b, respectively, lead to a translation in the $y$ direction and a vibration. The total of three vibrations we have generated is in accord with the $3n - 6$ rule. Since $v_2$ and $v_3$ are generated by partner GOs, their frequencies are degenerate.

If we now introduce a central atom Y into the molecule Z$_3$, Generator Table 5-6 acquires the added AV sets denoted by Y in Generator Table 5-7. Figure 5-22a depicts $v_1$. Easily visualized (see Figure 5-18g and h) are the $pz$-generated SMs which give rise to $T_z$ and the vibration $v_2$. We turn now to the three $px$- and $py$-generated SMs. From our discussion of N$_3^+$ we have already deduced the appearance of the SMs in the Z$y$ and Z$x$ AV equivalence sets generated by $px$ and $py$. In Figure 5-26 are shown the SMs of the

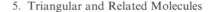

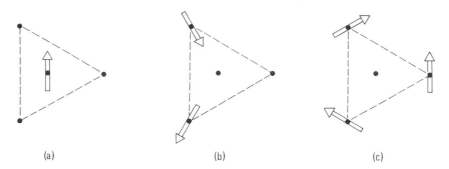

(a)                          (b)                          (c)

Figure 5-26. Symmetry motions of the $py$-generated atomic vectors in the (a) Y, (b) $Zy$, and (c) $Zx$ equivalence sets of trigonal planar $YZ_3$.

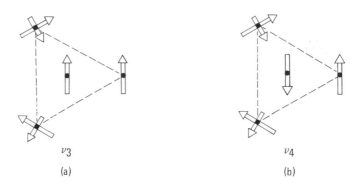

$\nu_3$                                   $\nu_4$

(a)                                       (b)

Figure 5-27. Linear combinations of the SMs shown in Figure 5-26a and c in which a small positive contribution of Figure 5-26b has been added.

$py$-generated AVs in the Y, $Zy$, and $Zx$ equivalence sets. From Generator Table 5-7 we note that there remain two non-vibrational motions, namely $T_y$ and $T_x$ which, of course, must be generated by $py$ and $px$, respectively. In the case of $T_y$, for example, it is clear that the appropriate linear combination of the AVs for this motion involves the sum in equal proportions of the SMs in Figure 5-26a and c plus a larger proportion of the negative of the SM in Figure 5-26b which serves to point the resultant motions of atoms B and C precisely along the $y$ direction. For the two vibrations of $py$ symmetry, imagine the Y atom moving as shown in Figure 5-26a. A counteracting vibratory motion would be obtained by adding either the peripheral atom motion shown in Figure 5-26b or the negative of the one in Figure 5-26c. The precise admixture of the motions in Figure 5-26b and c with that in Figure 5-26a is determined by the relative masses and force constants. These motions give rise to $\nu_3$ and $\nu_4$.

Similar considerations of the $px$-generated SMs in Table 5-7 give rise to $T_x$, $\nu_5$, and $\nu_6$. The total of six vibrational nodes is expected on the basis of the $3n - 6$ formula.

Table 5-8. A Useful Summary of Procedures for Generating Pictorial
Representations of Delocalized and Localized Bonding Views of a Molecule and
for Visualizing its Vibrational Modes

| Delocalized View | Localized View | Normal Modes |
|---|---|---|
| 1. Decide on best electron dash structure *using lowest possible formal charges and drawing resonance forms. In such forms some lp's may be involved in a π system.* | 1. Decide on which set(s) of doubly occupied delocalized MOs to localize. | 1. Identify the motional atomic vector (AV) equivalence sets and list them in the generator table. |
| 2. Establish the molecular geometry *from VSEPR theory of possible* | 2. Hybridize the **GO**s which give rise to the sets(s) in 1. | 2. Using **GO**s, sketch the symmetry motions (SMs). |
| 3. After preassigning lp's to (*s*) and suitably oriented (*p*) VAOs, identify all the VAO equivalence sets (*hybrids may also be used*) and list them in the generator table. | 3. Using the hybridized **GO**s, sketch the SOs. <br> 4. Sketch the MOs, being aware of the consequences of a central atom on which the full set of SOs (hybridized VAOs) is not generated by the hybridized **GO** set. *If necessary, rehybridize peripheral atom VAOs to locate bp's between atoms.* | 3. Fill in the rest of the generator table. (i.e., **GO**s and m entries). <br> 4. Take *appropriate* combinations of SMs generated by the same **GO** and sketch all the molecular motions. <br> 5. Identify the normal vibrational modes. |
| 4. Using **GO**s, sketch all the SOs, recalling that #SOs = #VAOs. | | |
| 5. Fill in rest of generator table (i.e., the necessary **GO**s and the *n*, *a*, or *b* entries). *Establish partner **GO** and SO sets.* | 5. Draw the resulting MO energy level diagram and occupy the appropriate levels with electrons. | |
| 6. Take linear combinations of SOs generated by the same **GO** and sketch all the MOs. | 6. Verify that the bond order per link is the same as in the delocalized view. | |
| 7. Draw an MO energy level diagram and occupy the appropriate levels with electrons. | | |
| 8. Deduce the bond order per link. | | |

# Summary

The new concepts introduced during our examination of the bonding in triangular and related molecules included: the determination of the coefficients of VAOs in SOs (e.g., in $H_3^+$) by means of simple trigonometric relationships, the recognition of partner sets of **GO**s and **MO**s, the preassignment of *sp*-out and *sp*-in hybrid VAOs on peripheral atoms in order to accommodate lp's (e.g., in $N_3^+$), the rehybridization of peripheral atom VAOs in order to localize bonding electron pairs between pairs of atoms (as in $N_3^+$), and the calculation of formal charges as an aid in deciding on the atom connections in molecules. In our discussion of $BF_3$, the importance of resonance and bond polarity was stressed as well as the role of $\pi$ bonding in molecular planarity [e.g., $B(OCH_3)_3$]. In visualizing the molecular motions of triangular planar $YZ_3$ species by means of **GO**s, a method was presented for determining the dominant SM contributions to the molecular motions when three SMs are generated by the same **GO**.

In Table 5-8 is an amplified version of Table 4-7. Additions are in italics.

PROBLEMS

1. For the species below, sketch electron dash structures indicating formal charges and generate delocalized and localized views of the bonding.

   (a) $Li_2H^+$; (b) $CH_3$; (c) $NO_3^-$; (d) $InCl_3$

2. Verify that no simpler **GO** than *f3″* gives an SO in Figure 5-9c.

3. Repeat Problem 1 including (*d*) VAOs on In for the $\pi$ system of $InCl_3$.

4. Repeat Problem 1 for the $BO_3$ portion of $B(OCH_3)_3$. Assume that the OC and CH bonds are localized.

5. Sketch the SMs in the *px* **GO** column of Table 5-7. Show how the $v_5$ and $v_6$ vibrations and the $T_x$ translation arise from these SMs.

6. $Fe(CO)_5$ is trigonal bipyramidal. Using sketches and an appropriate **GO** table, describe the motions of the trigonal planar $Fe(CO)_3$ moiety. (The FeCO bonds are all linear.)

# CHAPTER 6

# Bent Triatomic Molecules

Among the more common bent triatomic molecules are HCH, HOH, HSH, HSeH, HTeH, FMgF, ONO, ONO$^-$, O$_3$, FOF, I$_3{}^+$, FBrF, and OClO$^-$. In this chapter we treat two examples: one without $\pi$ bonding (H$_2$O) and one with $\pi$ bonding (NO$_2{}^-$).

## 6.1. H$_2$O

The electron dash structure of H$_2$O in Figure 6-1 tells us that there are four electron pairs around oxygen in a localized view. Four electron pairs can maximally repel one another by arranging themselves at the corners of a tetrahedron wherein the angles are 109°28′. As mentioned previously, we expect the bond angle in the chalcogen hydrides HChH (Ch = O, S, Se, Te) to be less than the tetrahedral angle since according to the VSEPR concept, lp–lp repulsions exceed bp–bp repulsions.

To develop the delocalized view of H$_2$O we begin by assigning the two oxygen lp's to the (O2s) and (O2px) VAOs in the axis system shown in Figure 6-2. The reason for choosing (O2px) here is that of the three (p) orbitals on oxygen, only (O2px) can be hybridized with (O2s) to provide localization for the oxygen lp's in (sp) hybrid VAOs which do not lie in the same plane as the O—H links. Another rationale for choosing (O2px) for an oxygen lone pair is that of all the oxygen (2p) VAOs, only (O2px) does not have the proper symmetry to interact with the (H1s) VAOs. Mixing (O2py) or (O2pz) with (O2s) would force the axis of the digonal lone pairs to be in the molecular plane and this is not expected on the basis of VSEPR theory. In Figure 6-2 it will be noted that the **GO** center (indicated by the triangle) is not coincident with the oxygen. This is because the oxygen is not the center of the molecule,

$$H - \overline{O} - H$$

Figure 6-1. Electron dash structure for $H_2O$.

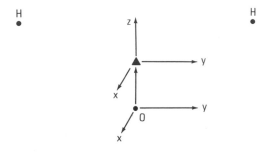

Figure 6-2. Axis system for $H_2O$. The **GO** center is denoted by the triangle.

Table 6-1.  Generator Table for
$H_2O$ (4 e$^-$)

| VAO Equivalence Sets | GOs | |
| --- | --- | --- |
| | $s$ | $py$ |
| $Opy = (O2py)$ | | n |
| $Opz = (O2pz)$ | n | |
| $Hs = (H_A1s), (H_B1s)$ | n | n |

since a point in the molecular plane directly below the oxygen in Figure 6-2, for example, does not feel the same potential as at the same distance directly above the oxygen in the molecular plane. It is clear that any **GO** location we decide upon should be equidistant from the two hydrogens and any point along the $z$ axis meets this requirement. We also want the **GO** center placed such that an as yet undetermined set of **GO** hybrids on it will point between the O and H atoms. This can be conveniently accomplished by locating the **GO** anywhere on the $z$ axis between the oxygen and a line connecting the two hydrogens. It is this type of reasoning which we will use to obtain our **GO** center in future examples. In Generator Table 6-1 are listed the three VAO equivalence sets we have available for bonding. The SOs are shown in Figure 6-3 and the linear combinations of SOs generated by the same GOs are depicted in Figure 6-4. The delocalized MO energy level diagram in Figure 6-5 yields the bond order of 1.0 expected on the basis of the electron dash structure.

The localized view of $H_2O$ is obtained as usual by using hybrids of the **GOs** associated with the occupied delocalized BMOs. Thus $spy_l$ and $spy_r$ on the **GO** center are directed toward the O—H interatomic regions where bond pair concentration is expected. If we were to hybridize the $(O2py)$ and $(O2pz)$

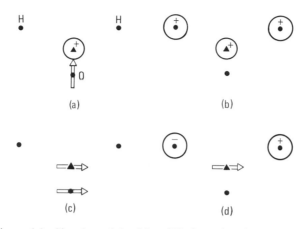

(a)                                              (b)

(c)                                              (d)

Figure 6-3. Sketches of the SOs of H$_2$O as given in Table 6-1.

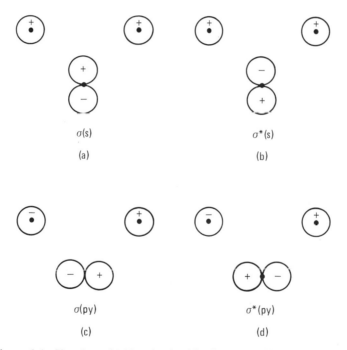

$\sigma(s)$                                      $\sigma^*(s)$

(a)                                              (b)

$\sigma(py)$                                     $\sigma^*(py)$

(c)                                              (d)

Figure 6-4. Sketches of MOs obtained by linear combinations of SOs.

VAOs such that they are rotated as shown in Figure 6-6, $spy_l$ would call in (H$_B$1$s$) and (O2$pz$), and $spy_r$ would call in (H$_A$1$s$) and (O2$py$). Each (H1$s$) VAO would interact with one (O2$p$) VAO to form a localized 2-center bond. Since we have no $s$-character in such a pair of rotated (O2$p$) VAOs, they would still be 90° apart and the HOH bond angle would also be 90°. While such a view is quite useful for H$_2$S, H$_2$Se, and H$_2$Te wherein the bond angles are

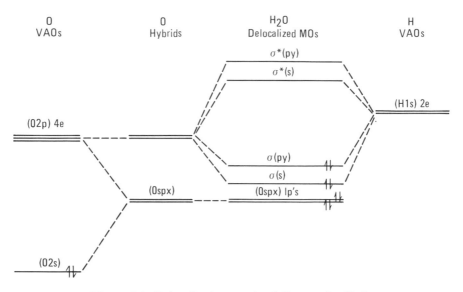

Figure 6-5. Delocalized energy level diagram for $H_2O$.

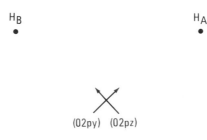

Figure 6-6. The result of rotating the $(O2py)$ and $(O2pz)$ VAOs from their positions as given in Figure 6-2 by means of hybridizing them.

92, 90, and 90°, respectively, it leaves something to be desired in the case of $H_2O$ for which the bond angle is 104°. By requiring less $p$-character in the oxygen lp's and more in the bp's, this objection can be overcome, however. Thus if the oxygen VAOs are nearly $sp^3$ hybridized, the HOH bond angle can be nicely accommodated. That is, the hybrids could be created with a bit more $s$-character in the lp's and a bit less in the bp's. This means, of course, that our original assumption that the $(O2s)$ and $(O2px)$ are reserved for lp's is modified somewhat. These conclusions are reflected in the localized MO energy level diagram in Figure 6-7 in which the oxygen VAOs are crudely assumed to be equivalent $sp^3$ orbitals.

An interesting point arises from a consideration of the relative energies of the lp's and the bond pairs in Figures 6-5 and 6-7. Photoelectron spectroscopy (PES) experiments confirm calculations that the first electron to be ionized

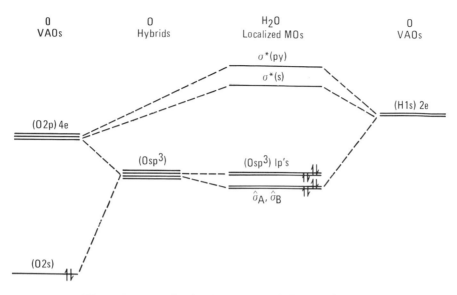

Figure 6-7. Localized MO energy level diagram for $H_2O$.

from water is from an oxygen lp. These results appear to conflict with the statement made in Chapter 1 that the delocalized view accommodates spectroscopic results better than the localized view. Recall, however, that in Figure 6-5 we localized the lp's. If we leave them in the canonical VAO's, the higher-energy lp is in ($O2px$), which indeed lies higher than the two BMOs. The ($O2s$) lp then lies below the two BMOs and this has also been verified by calculations and PES.

## 6.2. $NO_2^-$

From the electron dash structure of this molecular ion in Figure 6-8a we see that two resonance structures can be drawn, of which the average is represented by the electron dash structure in Figure 6-8b. It should be noted that in the axis system of Figure 6-9, the maximum delocalization of the $\pi$ bond in $NO_2^-$ suggested by Figure 6-8b requires the participation of the ($O2px$) VAO on each oxygen as well as the ($N2px$) VAO. This means that in addition to the $\pi$ bonding electron pair in Figure 6-8a, a lone pair from the oxygen associated with the other N—O link is also engaged in the delocalized $\pi$ system of Figure 6-8b. From these electron dash formulations we conclude that there is present a delocalized 3-center 2-electron $\pi$ bond and also a 3-center 2-electron non-bonding $\pi$ MO. We further can conclude from these electron dash structures and VSEPR considerations that $NO_2^-$ is bent. The experimentally determined bond angle of 115° suggests that the repulsion of the nitrogen lp by the combination of the two $\sigma$ bonds and the two half $\pi$ bonds is greater than the

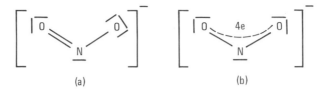

Figure 6-8. One resonance structure for $NO_2^-$ (a) and a representation of an average electron dash structure (b).

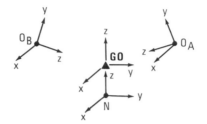

Figure 6-9. Axis system for $NO_2^-$.

Table 6-2. Generator Table for $NO_2^-$ (6 e$^-$)

| VAO Equivalence Sets | GOs | | | |
|---|---|---|---|---|
| | $s$ | $px$ | $py$ | $dxy$ |
| $N pr = (N2pz)$ | n | | | |
| $N pt = (N2py)$ | | | n | |
| $N p\pi = (N2px)$ | | n | | |
| $O pr = (O_B 2pz), (O_A 2pz)$ | n | | n | |
| $O p\pi = (O_B 2px), (O_A 2px)$ | | n | | n |

mutual repulsion of the $\sigma$ plus $\pi$ bonding electron density in the two links. As we will see shortly, the $\pi$ nonbonding electron pair is mainly localized on the oxygens and therefore these electrons are not repelled strongly by the nitrogen lp.

We begin our discussion of the delocalized view of the bonding in $NO_2^-$ by assigning lp's to VAOs. The oxygen lp's are assigned to the (O2s) and (O2py) VAOs on each oxygen in Figure 6-9 and the nitrogen lp is assigned to (N2s). We could, of course, localize these lp's further by placing them in a pair of digonal $spy$ VAOs on each oxygen. The MOs implied in Generator Table 6-2 are depicted in Figure 6-10. By reversing the sign of the nitrogen (or oxygen) SOs in Figure 6-10a–c, we obtain the antibonding counterparts of the BMOs shown. The delocalized MO energy level diagram is given in Figure 6-11 and the bond order of 1.5 agrees with that suggested by the electron dash structure.

The localized view is conveniently obtained by using $spy$ hybrid **GO**s to localize the two $\sigma$ bonding MOs, and $pxdxy$ hybrids to localize the occupied $\pi$

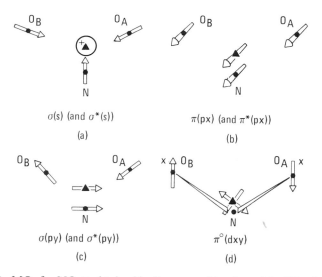

Figure 6-10. MOs for NO$_2$⁻ obtained by linear combination of the SOs given in Table 6-2 ((a)–(c)). The **GO**s giving rise to these MOs are also shown. The SO in (d) is a NBMO.

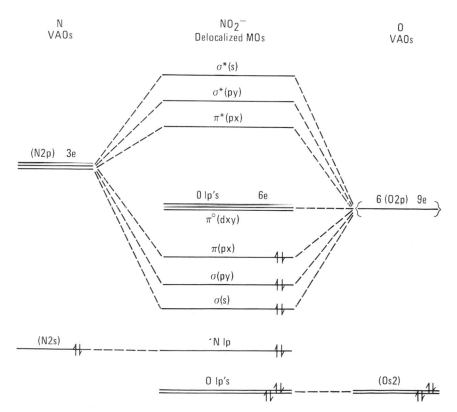

Figure 6-11. Delocalized MO energy level diagram for NO$_2$⁻.

Figure 6-12. Localized $\sigma$ MOs of $NO_2{}^-$ called in by digonal *spy* **GO**s.

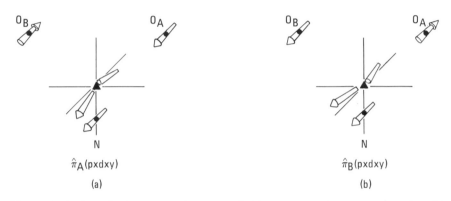

(a)                                                      (b)

Figure 6-13. Localized $\pi$ MOs of $NO_2{}^-$ called in by the $pxdxy_r$ (a) and $pxdxy_l$ (b) hybrid **GO**s.

MOs. The digonal *spy* hybrids call in contributions from the corresponding $(2pz)$ VAOs on the oxygens and $(N2py)$ and $(N2pz)$ VAOs which are suitably rotated via hybridization in order to localize electron density within the links (Figure 6-12). Increased localization of these MOs as well as of the nitrogen lp can be achieved by hybridizing the $(N2s)$, $(N2py)$, $(N2pz)$ set to nearly $sp^2$ VAOs. The $pxdxy$ hybrid **GO**s call in the $(O2px)$ VAOs as shown in Figure 6-13 but the nitrogen VAOs respond only with $(N2px)$. As we have come to appreciate in such cases, $\hat{\pi}_A(pxdxy)$ and $\hat{\pi}_B(pxdxy)$ include small antibonding contributions from one of the oxygen $(2px)$ VAOs in order to preserve orthogonality. These antibonding contributions also reduce the localized bond order of 1.0 to 0.5. In Figure 6-14 is depicted a localized MO energy level diagram.

The molecular motions of a bent $YZ_2$ molecule are summarized in Generator Table 6-3. For these motions, the **GO** center represents the center of mass of the molecule. It should be noted that while Generator Table 6-2 contains $s$ but not $pz$, the opposite obtains in Generator Table 6-3. This is because in the latter table the Z tangential AVs are included whereas the oxygen tangential AOs in $NO_2{}^-$ were preassigned lp's. Since $s$ generates SMs

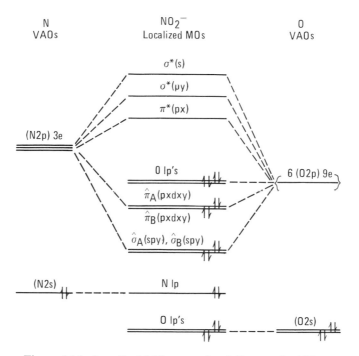

Figure 6-14. Localized MO energy level diagram for NO$_2^-$.

Table 6-3. Generator Table for Bent YZ$_2$

| | GOs | | | |
|---|---|---|---|---|
| AV Equivalence Sets | px | py | pz | dxy |
| Yr = Yz | | | m | |
| Yt = Yy | | m | | |
| Yπ = Yx | m | | | |
| Zr = Z$_A$z, Z$_B$z | | m | m | |
| Zt = Z$_A$y, Z$_B$y | | m | m | |
| Zπ = Z$_A$x, Z$_B$x | m | | | m |

with the Yr and Zr AV equivalence sets but $s$ does not generate a SM with Zt (whereas $pz$ does), we use $pz$ instead of $s$. Thus $pz$ generates more SMs than $s$. All the SMs are given in Figure 6-15. In Figure 6-15a we see that the $px$ **GO** calls in the Yx AV on nitrogen and $+Z_A x + Z_B x$ in the partner $Z_A x$, $Z_B x$ AV equivalence set on the Z atoms (see Generator Table 6-3). This linear combination, which is a sum of these two symmetry motions, gives rise to translation along $x$. By taking the difference between these symmetry motions (e.g., by reversing the sign of Yx), a rotation around the $y$ axis of the center of mass is obtained. In Figure 6-3b, $py$ is seen to call in $+Yy$ and the two

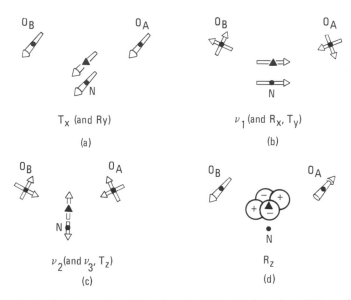

Figure 6-15. Sketches of the SMs given in Table 6-3 for a bent $YZ_2$ molecule.

partner sets on the Z atoms as $-Z_A y + Z_B y$ and $Z_A z - Z_B z$. Since we have three symmetry motions here, three molecular motions must be produced. One of these will shortly be seen to be a rotation and another a translation. Because one **GO** never gives rise to more than one rotation or one translation, the third motion must be a vibration. The molecular rotation around $x$ can be seen to arise from the linear combination $+Yy + Z_A y - Z_B y$ which represents the counterclockwise rotation of each atom at the same speed around the center of mass. In this molecular motion there is no contribution from $Z_A z, Z_B z$. The translation along the **GO** direction is easily visualized by sketching $Yy + a[-Z_A z + Z_B z] + b[-Z_A y + Z_B y]$, where the contribution of the middle term (given by $a$) exceeds that of $b$ in the third term so that the vector resultants are parallel to the **GO** (and to $Yy$). The precise ratio of the contributions (coefficients $a$ and $b$) of the terms associated with the $Z_A y, Z_B y$ and $Z_A z, Z_B z$ AV equivalence sets in this molecular motion depends on the bond angle. Only for an angle of 90° will it be 1 : 1. The third molecular motion is some linear combination of the symmetry motions that represent a vibration. A requirement of all vibrations is that the center of mass remain stationary and that there be no net angular momentum of the molecule (i.e., no net rotational motion) during any part of the vibration. Thus if N moves in the direction shown in Figure 6-3b, then we must have a contribution from $Z_A z - Z_B z$ to counteract the motion of Y. Depending on the relative masses and force constants there can be a contribution from $Z_A y$ and $Z_B y$ that directs the motions of these atoms either downward or upward from the mass center. A $+Z_A y - Z_B y$ contribution is ruled out, however, by the fact that it would

Table 6-4.  A Useful Summary of Procedures for Generating Pictorial
Representations of Delocalized and Localized Bonding Views of a Molecule and
for Visualizing its Vibrational Modes

| Delocalized View | Localized View | Normal Modes |
|---|---|---|
| 1. Decide on best electron dash structure using lowest possible formal charges and drawing resonance forms. In such forms, some lp's may be involved in a $\pi$ system | 1. Decide on which set(s) of doubly occupied delocalized MOs to localize. | 1. Identify the motional atomic vector (AV) equivalence sets and list them in the generator table. |
| 2. Establish the molecular geometry from VSEPR theory if possible. | 2. Hybridize the **GOs** which give rise to the sets(s) in 1. | 2. Using **GOs**, sketch the symmetry motions (SMs). |
| 3. After preassigning lp's to ($s$) and suitably oriented ($p$) VAOs, identify all the VAO equivalence sets (hybrids may also be used) and list them in the generator table. | 3. Using the hybridized **GOs**, sketch the SOs. | 3. Fill in the rest of the generator table. (i.e., **GOs** and m entries). |
| | 4. Sketch the MOs, being aware of the consequences of a central atom on which the full set of SOs (hybridized VAOs) is not generated by the hybridized **GO** set. If necessary, rehybridize peripheral atom VAOs to locate bp's between atoms. | 4. Take appropriate linear combinations of SMs generated by the same **GO** and sketch all the molecular motions. |
| 4. Using **GOs** *located at a molecular center or centroid*, sketch all the SOs, recalling that #SOs = #VAOs. | | 5. Identify the normal vibrational modes. |
| 5. Fill in rest of generator table (i.e., the necessary **GOs** and the $n$, $a$, or $b$ entries). Establish partner **GO** and SO sets. | 5. Draw the resulting MO energy level diagram and occupy the appropriate levels with electrons. | |
| 6. Take linear combinations of SOs generated by the same **GO** and sketch all the MOs. | 6. Verify that the bond order per link is the same as in the delocalized view. | |
| 7. Draw an MO energy level diagram and occupy the appropriate levels with electrons. | | |
| 8. Deduce the bond order per link. | | |

impart a rotation around $x$. The resulting vibration which can now be sketched is an unsymmetrical stretching mode ($v_3$ in the parlance of the spectroscopist).

In Figure 6-3c it is easy to see what combinations of the three symmetry motions depicted there give the translation along the $z$ direction of the center of mass. Since no combination gives a rotation, the other two combinations must be vibrations. Allowing Y to move in the direction shown in Figure 6-3c demands counteracting movements from the Z atoms with differing contributions of their two symmetry motions that depend upon the relative masses and force constants. Generally the relative contributions are such that one of the vibrations is a nearly pure bending mode, $v_2$, and the other a nearly pure bond stretching mode, $v_3$ (i.e., the symmetrical stretch called $v_1$ by spectroscopic convention). The rotation around $z$ of the center of mass is shown in Figure 6-3d. From the $3n - 6$ formula we expect three vibrational modes, as indeed we have found by employing **GO**s. Note that since none of the $p$ **GO**s are partners in the bent triatomic geometry, all the vibrations generated by them will have different frequencies.

# Summary

In examining $H_2O$ and $NO_2^-$ we became aware of the convenience of placing our **GO** center between the middle atom and the pair of atoms bound to it on a line which bisects the angle in the molecular plane. In the localized view of $H_2O$ we saw how VAOs on the oxygen could be hybridized to form highly localized 2-center 2-electron bonds to the hydrogens. A feature of $NO_2^-$ is the possibility of resonance, which leads to a delocalized $\pi$ system. Finally, the molecular motions of bent $YZ_2$ molecules were elucidated by the **GO** method.

An updated recipe for using the **GO** approach to bonding and molecular motions appears in Table 6-4.

PROBLEMS

1. Draw electron dash structures indicating formal charges and develop the delocalized and localized bonding views of (a) $O_3$; (b) $H_2Te$ (90° angle); (c) $OF_2$.

2. From their delocalized MO energy level diagrams, rationalize why the bond distance in NO decreases when it is ionized to $NO^+$ but the bond lengths are not significantly changed on going from $NO_2^-$ to $NO_2$.

3. Rationalize why the bond angle in $NO_2^-$ (115°) increases to 134° in $NO_2$.

4. When a positive species, such as a proton reacts with water, it attacks the middle atom in the molecule but when such a reaction occurs with $NO_2^-$, one of the end atoms is attacked. Account for these phenomena by considering the delocalized bonding views of these substrates.

# CHAPTER 7

# Polygonal Molecules

Examples of ring compounds in organic and inorganic chemistry are legion. Rings having more than three atoms and which are saturated generally possess a puckered ring structure with roughly tetrahedral bond angles, in accord with VSEPR considerations. The members of such rings are connected by $\sigma$ bonds only and each member has in addition two lone pairs (e.g., $S_8$), a lp and bp (e.g., $c$-$(PPh)_5$), or two bp's [e.g., $c$-$C_6H_{12}$]. The bonding in saturated systems can be qualitatively understood by treating each three-atom fragment as a bent triatomic molecule. An important class of ring molecules is that of planar species. Several members of this polygonal class are particularly interesting from a bonding point of view. We have already examined one example of such a molecule, namely, $N_3^+$. We now address ourselves to ring systems with more than three members, which lie in the same plane *only* because of the presence of delocalized $\pi$ bonding.

## 7.1. $Te_4^{2+}$

Four different electron dash structures (Figure 7-1) can be drawn for this deep purple cation. The acyclic structure in Figure 7-1a is unreasonable in view of the adjacent charges. As implied by one of the four resonance structures in Figure 7-1b, however, both the charges and the $\pi$ bonding are more effectively delocalized around a cyclic four-membered ring. The average electron dash structure for these resonance forms is shown in Figure 7-1c. Although the structure in Figure 7-1d is also cyclic, the $\pi$ bond in it is restricted to the position shown since migration of the double bond to any other position would bring the nonplanar trisubstituted Te atom into a planar configuration with its three substituents (Figure 7-1e). *It is a cardinal rule in drawing*

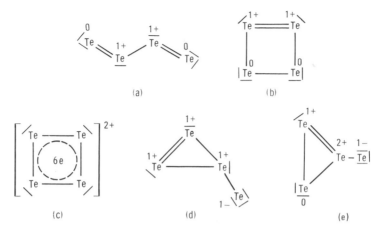

Figure 7-1. Possible electron dash structures for $Te_4^{2+}$ [(a), (b), (d) (e)]. In (c) is shown an average electron dash structure of the resonance forms exemplified by the structure in (b).

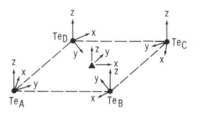

Figure 7-2. Axis system for $Te_4^{2+}$.

*resonance structures that only electrons may be shifted and not atoms.* Figure 7-1d and 7-1e depict electron dash structures that are not resonance structures. Both of them also suffer from the destabilizing presence of adjacent like charges. From Figure 7-1b and c, we see that in order for delocalization of the π bond to occur, two lp's from the lower two Te atoms in the resonance structure in Figure 7-1b must be housed in nonbonding π MOs. The $Te_4^{2+}$ ring is classified as *aromatic* since it obeys the $4n + 2$ rule. This rule says that a ring is aromatic if the total number of π electrons is equal to $4n + 2$ where $n$ is an integer. In $Te_4^{2+}$, $n = 1$

For the delocalized view of $Te_4^{2+}$ we first assign the lp's to (Te5s) VAOs. Using the axis system in Figure 7-2, we construct Generator Table 7-1. The SOs generated from the various **GO**s are depicted in Figure 7-3. These SOs are the MOs of the ring and their bonding, antibonding, and nonbonding natures become evident from examining their nodal patterns. The nodal patterns for the MOs in the LCAOs implied in Figure 7-3b and c are shown in Figure 7-4a, b and Figure 7-4c, d respectively. Here we assume in determining the overall bonding nature of an MO that σ interactions outweigh the in-plane interactions of the Te*pt* equivalence set. The MOs designated as being of the π type

Table 7-1. Generator Table for Te$_4$$^{2+}$ (14 e$^-$)

| VAO Equivalence Sets | GOs | | | | | | | | | |
|---|---|---|---|---|---|---|---|---|---|---|
| | $s$ | $px$ | $py$ | $pz$ | $dxy$ | $dx^2-y^2$ | $dxz$ | $dyz$ | $f2''$ | $g4''$ |
| Tep$\sigma$ = (Te$_A$5$py$), (Te$_B$5$py$), (Te$_C$5$py$), (Te$_D$5$py$) | b | n | n | | a | | | | | |
| Tep$\pi$ = (Te$_A$5$pz$), (Te$_B$5$pz$), (Te$_C$5$pz$), (Te$_D$5$pz$) | | | | b | | | n | n | a | |
| Tept = (Te$_A$5$px$), (Te$_B$5$px$), (Te$_C$5$px$), (Te$_D$5$px$) | | n | n | | | b | | | | a |

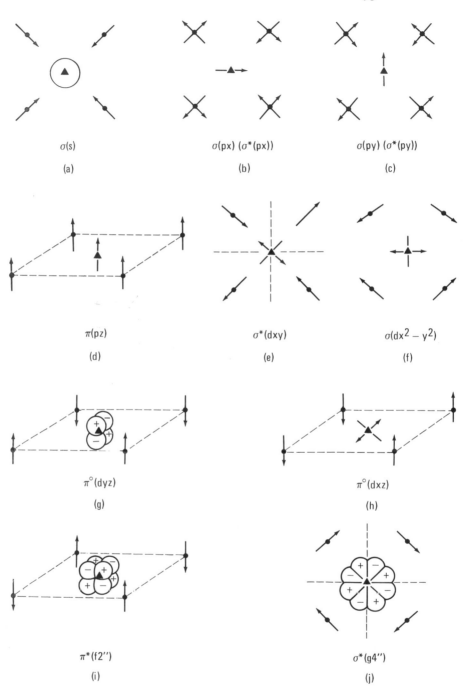

Figure 7-3. Sketches of the SOs of $Te_4^{2+}$ in Table 7-1. In this and future figures the pictorial representations of $p$ and $d$ orbitals will often be given by two-dimensional drawings instead of three-dimensional shafts.

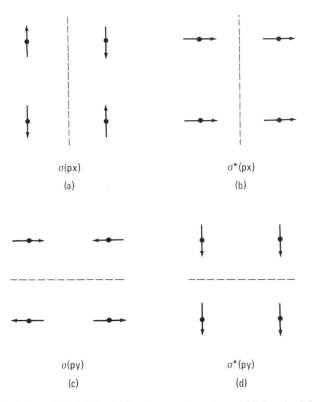

$\sigma(px)$ (a)          $\sigma^*(px)$ (b)

$\sigma(py)$ (c)          $\sigma^*(py)$ (d)

Figure 7-4. Sketches of the MOs obtained upon linearly combining the SOs in Figure 7-3b [(a), (b)] and in Figure 7-3c [(c), (d)].

maintain their orientation perpendicular to the plane of the molecule as in $N_3^+$. The MOs labeled as $\sigma$ MOs are in the plane of the molecule as they were in $N_3^+$. The delocalized MO energy level diagram in Figure 7-5 implies that the average bond order is 5/4 per link. Morever, since there are six electrons in the $\pi$ MOs, we have agreement with the $4n + 2$ aromaticity rule (i.e., $4n + 2 = 6$ if $n = 1$) and this $\pi$ occupation is also reflected in the average electron dash diagram in Figure 7-1c.

The axis system chosen for $Te_4^{2+}$ appears unusual in that contrary to the axis system used for $N_3^+$ (Chapter 5) and for further examples of polygonal species in this chapter, neither the $x$ nor the $y$ axis points to an atom. Because of the fourfold symmetry of a square as well as of a pair of orthogonal axes, there is no overriding reason to point the axes toward an atom. This is because the **GO**s are easy to identify (i.e., no hybrid **GO**s are necessary) whether the $x$ and $y$ axes point to atoms or to the midpoints between them. The reason we chose to point these axes between the atoms is that the MOs are more delocalized in the delocalized view than if we had chosen to direct $x$ and $y$ toward the atoms (see Problem 8).

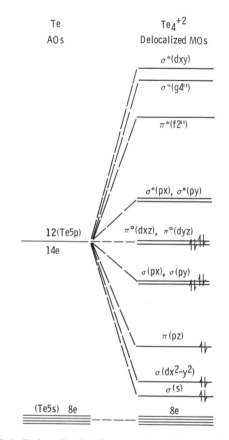

Figure 7-5. Delocalized MO energy level diagram for $Te_4{}^{2+}$.

For the localized view let us hybridize the GOs of the $\sigma$ MO set separately from those of the $\pi$ MO set. The main lobes of an $pxpydx^2 - y^2$ hybrid set for the $\sigma$ MOs point to the edges of the square in Figure 7-2. These lobes call in 2-electron $\sigma$ bonds between each pair of tellurium atoms, which involve linear combinations of the in-plane radial and tangential VAOs. Neglecting neighbor overlap, we can reorient these orthogonal in-plane ($p$) VAOs as shown in Figure 7-6. For the localized $\pi$ bonds a $pzdxzdyz$ GO set is required. The main lobes of each member of such a set point as shown in Figure 7-7a. Together the three members point to the corners of a trigonal prism. The orientation of these hybrids with respect to rotation about the $z$ axis of $Te_4{}^{2+}$ is arbitrary and two possible directions for the main lobes of these hybrids will be discussed. In one of these orientations one member hybrid points its main lobes over and under an edge of the molecule, and in the other these two lobes are pointed over and under an atom center. Although the localized $\pi$ MOs are all of equal energy, they do not all have identical shapes (Figure 7-7b and c) and hence

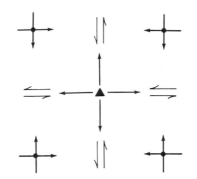

Figure 7-6. Localized bonds in $Te_4^{2+}$ called in among the reoriented Te ($p$) VAOs by the $pxpydx^2 - y^2$ **GO** lobes.

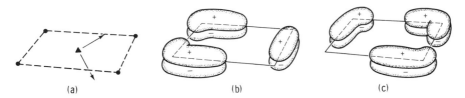

Figure 7-7. Main lobes of one member of a $pzdxzdyz$ **GO** set (a). In (b) and (c) are shown two possible orientations of the three localized $\pi$ MOs in $Te_4^{2+}$.

different atom contributions are involved. The localized MO energy level diagram is easy to draw and this is left as an exercise at the end of the chapter.

The molecular motions of a square tetra-atomic molecule such as $Te_4^{2+}$ is summarized in Generator Table 7-2. Note that these SMs look like the MOs in Generator Table 7-1 and in Figures 7-3 and 7-4. The SMs corresponding to these figures are $v_1$ (Figure 7-3a), $v_2$ (Figure 7-4a), $T_x$ (Figure 7-4b), $v_3$ (Figure 7-4c), $T_y$ (Figure 7-4d), $T_z$ (Figure 7-3d), $v_4$ (Figure 7-3e), $v_5$ (Figure 7-3f), $R_x$ (Figure 7-3g), $R_y$ (Figure 7-3h), $v_6$ (Figure 7-3i), and $R_z$ (Figure 7-3j). The six frequencies found in our **GO** analysis is in agreement with the $3n - 6$ rule discussed earlier.

## 7.2. $c$-$C_4H_4$

Cyclobutadiene is an unstable four-membered ring which has recently been detected in a low-temperature photochemical reaction. The $\pi$ delocalization in this molecule is similar to that encountered in $Te_4^{2+}$ but the present molecule is not aromatic. As we shall see, the electronic structure of the ground state depends upon whether the molecule is square or rectangular. For the square geometry the ground state is a triplet (two unpaired electrons) and, hence,

Table 7-2. Generator Table for $Z_4$

| AV Equivalence Sets | GOs | | | | | | | | | |
|---|---|---|---|---|---|---|---|---|---|---|
| | $s$ | $px$ | $py$ | $pz$ | $dxy$ | $dx^2 - y^2$ | $dxz$ | $dyz$ | $f2''$ | $g4''$ |
| $Zr$ | m | m | m | | m | | | | | |
| $Zt$ | | m | m | | | m | m | | | m |
| $Z\pi$ | | | | m | | | m | m | m | |

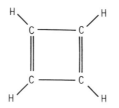

Figure 7-8. Electron dash structure for cyclobutadiene.

paramagnetic. For the rectangular geometry the ground state can be a singlet (all spins paired) and, hence, diamagnetic. Which structure applies to the species of minimum energy has not yet been settled. We first discuss the square geometry.

The $\sigma$ bonding system of $c$-$C_4H_4$ is analogous to that of $Te_4^{2+}$. The two species are related to each other in the same manner as the $\sigma$ systems of $N_3^+$ and $c$-$C_3H_3^+$ are related. In the present discussion the $\pi$ system is the feature of interest. We will discuss the $\sigma$ system of a planar $c$-$C_nH_n^x$ system in our examination of $c$-$C_5H_5^-$ in the next section. In our discussion of the $\pi$ MOs of square $c$-$C_4H_4$, we will see that these MOs are qualitatively the same as those of $Te_4^{2+}$. The essential difference between the two molecules is that the $\pi$ system of $c$-$C_4H_4$ has two electrons less than that of $Te_4^{2+}$ (Figure 7-8). Consequently there are only two electrons available for the degenerate NBMOs $\pi^0(dxz)$, $\pi^0(dyz)$. Hence one electron will go in each NBMO and the two spins will be parallel to give a triplet ground state.

Regarding localization, we recall that only MOs with the same occupation numbers can be mixed to obtain localized MOs. Hence no further localization is possible for the one doubly occupied $\pi$ MO. The two singly occupied MOs, on the other hand, are degenerate partners and mixing them will not produce localized MOs, but only a different set of degenerate partners. Thus no really localized view is possible.

Thinking in terms of resonance structures, one would be led to a description of the $\pi$-electronic structure of $c$-$C_4H_4$ as an equal mixture of the two structures implied by Figure 7-8. However, each of these resonance structures involves doubly occupied orbitals only, and hence each corresponds to a singlet state. It follows that this is also true of any linear combination of the two structures. Hence the resonance description is appropriate to the lowest-lying singlet and not the triplet ground state. This argument holds for all molecules with resonance structures. Thus, in general, the resonance description corresponds to the lowest singlet state of the molecule and not to its ground state if the ground state happens to be of higher multiplicity than a singlet.

The rectangular geometry of $c$-$C_4H_4$ can be obtained by deforming the square geometry into a rectangle through elongation along the $x$ axis of the **GO** center. (It is of course arbitrary whether the elongation occurs along the $x$ or the $y$ axis.) As a result of the distortion, $\pi(pz)$ becomes less bonding,

$\pi^0(dxz)$ gains bonding character to become $\pi(dxz)$, $\pi^0(dyz)$ gains antibonding character to become $\pi^*(dyz)$, and $\pi^*(f2'')$ reduces its antibonding character. These can be seen from the changes in the pairwise interactions among the MO lobes. If the distortion and, hence, the removal of the degeneracy in the nonbonding $\pi$ MOs is sufficiently large, then the ground state will correspond to double occupancy of $\pi(dxz)$ with $\pi^*(dyz)$ being empty, so that a singlet state results. Molecules frequently distort geometrically in order to remove an orbital degeneracy, and a sufficient orbital energy separation in such a distortion leads to electron pairing in the lower-lying MO. This process is called the "Jahn–Teller effect," in honor of its discoverers.

The singlet $\pi$ MO state lends itself to localization since the fully occupied delocalized $\pi$ MOs come from a $pz$ and a $dxz$ **GO**. This $pd$ hybrid set consists of an oppositely directed pair of hybrid **GOs** (such as were used for one of the localized forms of $CO_2$) pointing towards the midpoints of the shorter sides of the rectangle. They generate localized MOs which provide 2-center 2-electron bonds between the corresponding atoms. Thus, the localized bond picture corresponds to the electron dash structure of Figure 7-8 having the two double bonds between those atoms which, *in the rectangular geometry*, lie closer to each other.

The vibrational modes of square $c\text{-}C_4H_4$ can be obtained by generating the molecular motions of the $C_4$ square and the $H_4$ square separately, using the **GO** center to also represent the center of mass, and then taking linear combinations of those motions generated by the same **GO**. Of course, the motions of the $C_4$ square duplicate those of the $Z_4$ square discussed in the previous section. By following along in Figures 7-3 and 7-4 it can be seen that linear combinations of duplicates of the SMs shown in each figure lead to $v_1$, $v_2$ (Figure 7-3a); $v_3$, $v_4$ (Figure 7-4a); $v_5$, $T_x$ (Figure 7-4b); $v_6$, $v_7$ (Figure 7-4c); $v_8$, $T_y$ (Figure 7-4d); $v_9$, $T_z$ (Figure 7-3d); $v_{10}$, $v_{11}$ (Figure 7-3e); $v_{12}$, $v_{13}$ (Figure 7-3f); $v_{14}$, $R_x$ (Figure 7-3g); $v_{15}$, $R_y$ (Figure 7-3h); $v_{16}$, $v_{17}$ (Figure 7-3i) and $v_{18}$, $R_z$ (Figure 7-3j). The $3n - 6$ formula tells us that we should have eighteen vibrations and indeed our GO analysis verifies this.

# 7.3. $c\text{-}C_5H_5{}^-$

The cyclopentadienide ion, made by proton abstraction from cyclopentadiene, is an important intermediate in the synthesis of a wide range of organometallic compounds. From the electron dash structure in Figure 7-9, it can be seen that part of the stability of this anion arises from its aromatic character. Thus each carbon contributes one $\pi$ electron and the carbon bearing the negative charge in a resonance structure such as seen in Figure 7-9a contributes one more. Again $n = 1$ in the $4n + 2$ rule. In discussing the bonding of $c\text{-}C_5H_5{}^-$ as well as planar aromatic species of the general formula

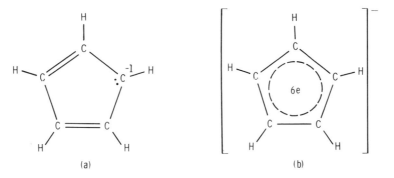

Figure 7-9. An electron dash structure of $c$-$C_5H_5^-$ (a). In (b) is shown an average electron dash structure of the resonance forms exemplified by (a).

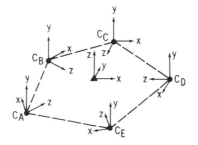

Figure 7-10. Axis system for $c$-$C_5H_5^-$.

$c$-$C_nH_n^x$, it is convenient to assign the H—C electron pairs in the Lewis structure to localized MOs. It should be recalled from our discussion of the delocalized view of $N_3^+$ that the nitrogen lone pairs could be localized in (N$spo$) VAOs. Thus it was not necessary to choose canonical AOs to form the VAO equivalence sets. Any hybrid AO set could serve equally well. In $N_3^+$ we used an inwardly directed (N$spi$) VAO [having less $s$-character] along with (N$2px$) and (N$2py$) for the delocalized bonding. In $c$-$C_nH_n^x$ [or any cyclic planar $c$-$(ZY)_n^x$ species for that matter] we can follow a similar path. In the case at hand, we use (C$spo$) VAOs and corresponding (H$1s$) VAOs for localization of the C—H bonds and we employ (C$spi$) VAOs for the radial $\sigma$ bonding. Though the outward- and inward-pointing hybrids have different admixtures of ($s$) and ($p$) character, we will not be concerned with these differences here.

Employing the axis system in Figure 7-10 we can construct Generator Table 7-3. In arriving at the proper GOs in this table it is of great importance to appreciate the fact that $p$, $d$, and $f$ GOs contained in it have "dumbbell," "four leaf clover," and cube-like symmetries, respectively, and that these symmetries

Table 7-3. Generator Table for $C_5$ framework in $c\text{-}C_5H_5^-$ (16 e⁻)

| VAO Equivalence Sets | GOs | | | | | | | | | | |
|---|---|---|---|---|---|---|---|---|---|---|---|
| | $s$ | $px$ | $py$ | $pz$ | $dxy$ | $dx^2 - y^2$ | $dxz$ | $dyz$ | $f2'$ | $f2''$ | $h5'$ |
| Csp-in = $(C_A spi)$, $(C_B spi)$, $(C_C spi)$, $(C_D spi)$, $(C_E spi)$ | b | b | b | | a | a | | | | | a |
| Cpx$\sigma$ = $(C_A 2px)$, $(C_B 2px)$, $(C_C 2px)$, $(C_D 2px)$, $(C_E 2px)$ | | a | a | | b | b | | | | | |
| Cpy$\pi$ = $(C_A 2py)$, $(C_B 2py)$, $(C_C 2py)$, $(C_D 2py)$, $(C_E 2py)$ | | | | b | | | b | b | a | a | |

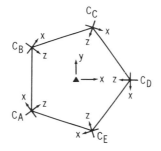

Figure 7-11. Top view of the orbital and **GO** in-plane axes for $c$-$C_5H_5^-$.

are super-imposed on the pentagonal geometry of the five carbons of $c$-$C_5H_5^-$. Using the schematic top view of the system shown in Figure 7-11 as an aid, it is seen, for example, that a $px$ **GO** calls in ($C_C spi$). ($C_D spi$), and ($C_E spi$) with positive signs in the hybrid main lobes while the same VAOs of $C_A$ and $C_B$ are called in with negative signs. Moreover, the coefficients for these VAOs will be in the order $C_D > C_A = C_B > C_C = C_E$. As a second example, it is easy to see from Figure 7-11 that the $dx^2 - y^2$ **GO** calls in ($sp_r$) lobes on $C_A$, $C_B$, and $C_D$ with the same sign but that these VAOs on $C_C$ and $C_E$ are called in with opposite signs. From a consideration of each SO in Figure 7-12, their overall bonding or antibonding nature indicated in Generator Table 7-3 can be determined. Because of the relative complexity of the molecule, no attempt has been made in Figure 7-12 to indicate the VAO coefficients by relative sizes of the arrows. This less rigorous procedure will be followed generally throughout the remainder of this book.

Linear combinations of the pairs of SOs generated by the same **GO**s give rise to pairs of MOs which are split apart in energy. Because of the fivefold symmetry of pentagonal $C_5H_5^-$, the determination of which linear combination in Figure 7-13 corresponds to the BMO and the ABMO is difficult. We would expect that the BMO consists mainly of the dominantly bonding SO generated in the $Csp$-in equivalence set and that the ABMO would consist mainly of the dominantly antibonding SO generated in the $Cpx\sigma$ equivalence set. Similar considerations hold for the remaining MOs. Since ten of the twenty-six valence electrons are accounted for in the localized C—H bonds, sixteen electrons remain for filling the eight BMOs in Figure 7-14. Of course, the order of the BMOs and ABMOs in this diagram is not possible to ascertain without an accurate calculation. The average $\sigma$ bond order is 1.0 and the average $\pi$ bond order is $3/5$, giving an overall average bond order of $8/5$ per link. As expected from the $4n + 2$ aromaticity rule ($n = 1$), there are six electrons in the $\pi$ MO system.

It should be noted that the tangential ($px$) VAOs are given the set label $Cpx\sigma$ because their overlap resembles $\sigma$ interactions, albeit at an angle. In

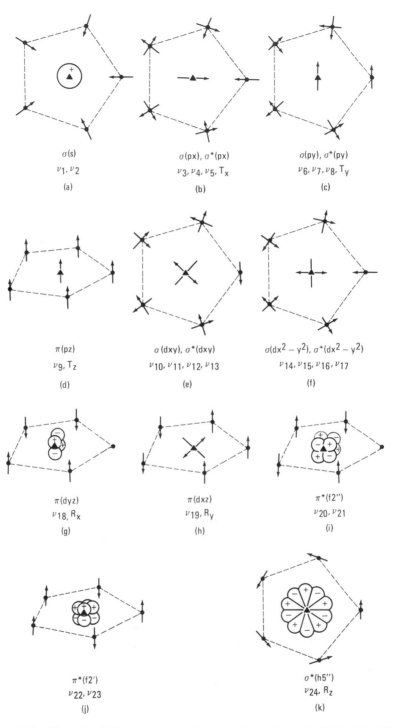

Figure 7-12. SOs of $c\text{-}C_5H_5{}^-$ corresponding to those given in Table 7-3. These sketches also represent the SMs of a pentagonal array of atoms.

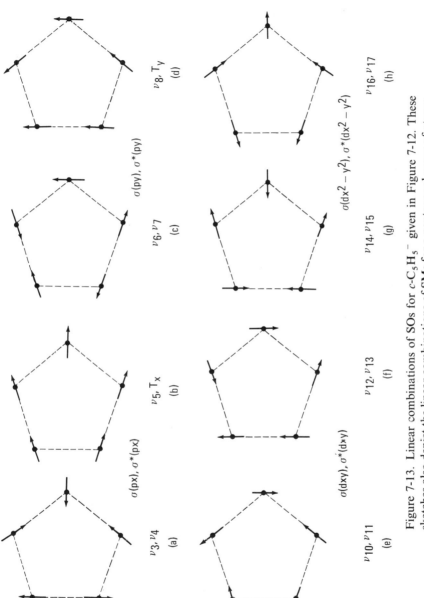

Figure 7-13. Linear combinations of SOs for $c$-$C_5H_5^-$ given in Figure 7-12. These sketches also depict the linear combinations of SMs for a pentagonal array of atoms.

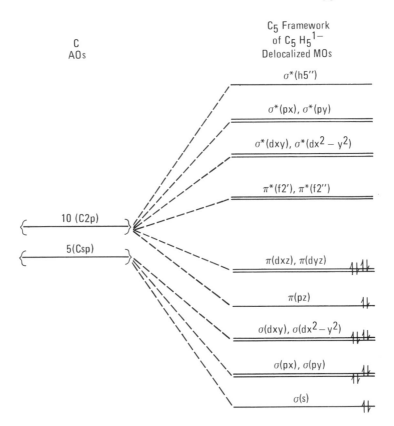

Figure 7-14. Delocalized MO energy level diagram for $C_5$ framework of $c\text{-}C_5H_5^-$.

higher-order polygons such overlaps become increasingly "head on" and hence more $\sigma$ in their appearance. In $Te_4^{2+}$ these VAOs were given the set table label $Tept$ (where t denotes tangential) since in a square system such orbitals are exactly halfway between $\sigma$- and $\pi$-type overlap. Note also that we consistently refer to MOs composed of inwardly directed VAO lobes in polygonal molecules as $\sigma$-type MOs even though they also involve angular overlap. Actually these MO labels apply in a strict sense only to diatomics. In more complex systems these designations are more arbitrary. We will adopt the convention that MOs composed of ($p$) VAOs which interact in a truly "side-on" manner will be labeled $\pi$ MOs, whereas less than strictly side-on overlap of these VAOs will lead to MOs labeled $\sigma$.

The $\sigma$ bonding MOs in $c\text{-}C_5H_5^-$ stem from an $sp^2d^2$ hybrid **GO** set whose members point their main lobes toward the corners of a regular pentagon. By directing the lobes midway between the carbon atoms, we conclude that the five localized $\sigma$ bond pairs can be housed in five 2-center bonds, each of which could be made up of C2$p$ VAOs from neighboring carbon atoms, as shown

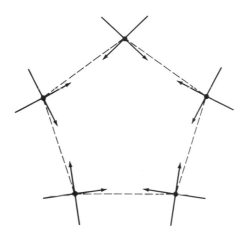

Figure 7-15. Carbon VAOs in $c$-$C_5H_5^-$ reoriented for localized bonds between C atoms.

in Figure 7-15. Taking advantage of the partially hybridized carbon **VAOs** we already have, however, we could hybridize C2s with *both* the C2px and the C2pz **VAOs** so that we have a new Cspo, and two C2$sp^y$ hybrids directed along the edges of the pentagon. Localizing the occupied $\pi$ MOs of Figure 7-14 involves the use of a $pd^2$ **GO** hybrid which we have already encountered in Te$_4^{2+}$. The directions we choose for the trigonally directed members of this set are arbitrary and in Figure 7-16 are shown two orientations of the localized $\pi$ bonding electron pair regions which result from two convenient orientations of the $pd^2$ hybrid **GOs**. The top view in Figure 7-16a shows that two of these localized MOs are equivalently positioned with respect to atom centers and consequently have the same shape. The third localized MO, being differently placed with respect to atom centers, has a different shape. This figure shows that the Lewis structure in Figure 7-9a, which implies two double bonds and a doubly occupied ($p$) AO, is a somewhat misleading oversimplification in that the unique localized MO is almost a 3-center bond and the double bonds are correspondingly polarized.

Since the choice of the carbon toward which we pointed the first hybrid **GO** is arbitrary, there are five different localized MO *sets*. These five MO representations are not to be superimposed (as resonance structures would be). Rather, they are all equivalent and each by itself is an adequate representation of the molecular state.

Another useful localized description for $c$-$C_5H_5^-$ is obtained from a different orientation of the hybrid **GOs**. Pointing one of these hybrid **GOs** toward the midpoint of a bond gives rise to the localized MOs shown in Figure 7-16c and d. It is apparent that while all these **GOs** point between carbon atom pairs, two are off-center. The localized MO energy level diagram is easy to draw and this is left for the reader as an exercise.

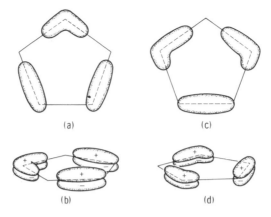

Figure 7-16. Two views of one orientation of the localized $\pi$ MOs in $c$-$C_5H_5^-$ [(a), (b)]. Two views of a second orientation are shown in (c) and (d).

We thus find that the three $\pi$ electron pairs in $c$-$C_5H_5^-$ can be considered as housed in three localized regions toward which the $pd^2$ hybrid **GO**s point. It may be mentioned that the approximately symmetrical distribution of $\pi$ electron pairs around any planar ring is a function only of the number of such pairs and is independent of the number of atoms making up the ring.

The molecular motions of $C_5H_5^-$ can also be obtained from Figures 7-12 and 7-13. These motions are linear combinations of the SMs of the carbons and the hydrogens which these figures represent.

## 7.4. $c$-$C_6H_6$, $c$-$C_7H_7^+$, $c$-$C_8H_8^{2-}$

The average electron dash structures of benzene, the tropylium ion, and the cyclooctatetraenide ion shown in Figure 7-17 tell us that all of these species are aromatic. That is, in the $4n + 2$ rule $n = 1$ for $c$-$C_6H_6$ and $c$-$C_7H_7^+$ while in $c$-$C_8H_8^{2-}$ $n = 2$. Since the bonding treatment of these molecules using the axis systems in Figure 7-18 is entirely analogous to that of $c$-$C_5H_5^-$, the symmetry orbitals and motions in Figures 7-19 to 7-21 generated from the AO and AV equivalence sets in Generator Tables 7-4 to 7-6 are given without further comment. In Figures 7-19 to 7-21 no attempt has been made to show relative sizes of orbitals or to indicate hybrid character in the (C2$spi$) VAOs. When drawing the SOs or SMs generated by the various **GO**s, care must be taken to observe the geometric relationships between given AOs or AVs, respectively, and the **GO**. The drawing of the delocalized MO energy level diagrams are left as an exercise.

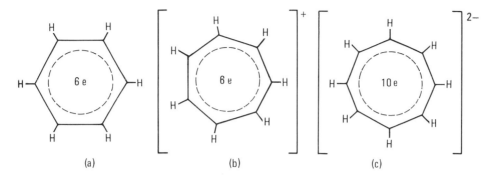

Figure 7-17. Average electron dash structures for $c$-$C_6H_6$ (a), $c$-$C_7H_7^+$ (b), and $c$-$C_8H_8^{2-}$ (c).

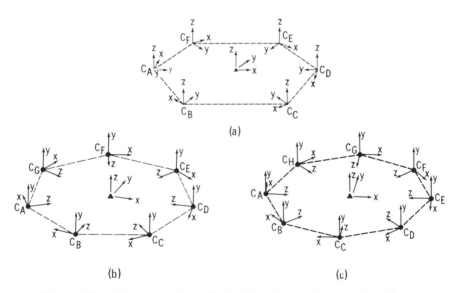

Figure 7-18   Axis systems for $c$-$C_6H_6$ (a), $c$-$C_7H_7^+$ (b), and $c$-$C_8H_8^{2-}$ (c).

## Summary

In this chapter we gained an appreciation for delocalized and localized bonding systems in planar ring-like molecules such as $Te_4^{2+}$, $c$-$C_4H_4$, $c$-$C_5H_5^-$, $c$-$C_6H_6$, $c$-$C_7H_7^+$, and $c$-$C_8H_8^{2-}$. All of these examples are aromatic (i.e., they obey the $4n + 2$ rule) except $c$-$C_4H_4$. In studying the latter molecule, the importance of Jahn–Teller distortions in removing orbital degeneracy was brought out.

In Table 7-7 is contained our latest amendment to the recipe for the **GO** approach.

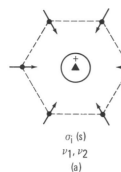
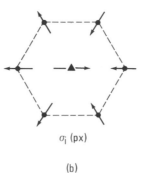
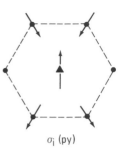

$\sigma_i$ (s)

$\nu_1, \nu_2$

(a)

$\sigma_i$ (px)

(b)

$\sigma_i$ (py)

(c)

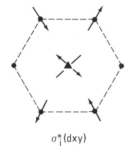
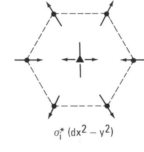
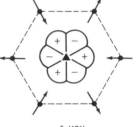

$\sigma_i^*$ (dxy)

(d)

$\sigma_i^*$ (dx$^2$ − y$^2$)

(e)

$\sigma_i^*$ (f3′)

$\nu_{17}, \nu_{18}$

(f)

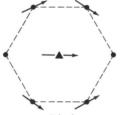
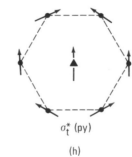

$\sigma_t^*$ (px)

(g)

$\sigma_t^*$ (py)

(h)

$\sigma_t$ (dxy)

(i)

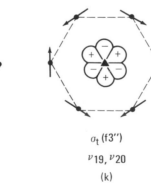
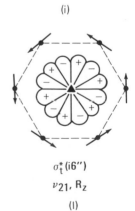

$\sigma_t$ (dx$^2$ − y$^2$)

(j)

$\sigma_t$ (f3″)

$\nu_{19}, \nu_{20}$

(k)

$\sigma_t^*$ (i6″)

$\nu_{21}, R_z$

(l)

Figure 7-19. Sketches of SOs of $c$-C$_6$H$_6$ and their linear combinations as given in Table 7-4. These sketches also represent the SMs of a hexagonal array of atoms. The subscripts i and t refer to "inward" pointing and "tangential" $\sigma$ MOs. Many of the molecular motions are not labeled since they arise from linear combinations of SMs generated by the same **GO** as in (s) and (t).

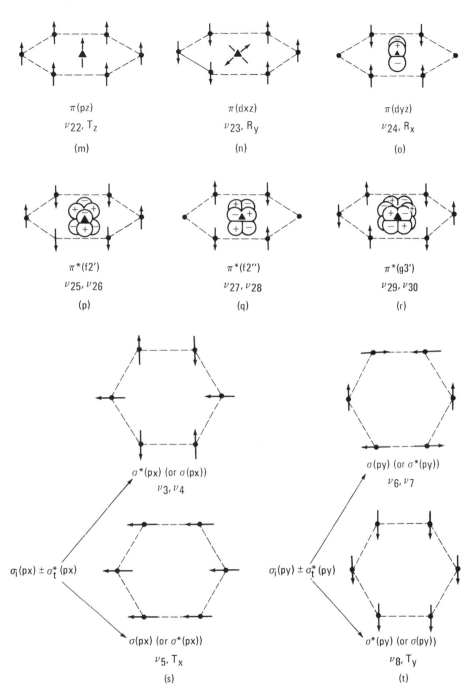

Figure 7-19 (continued)

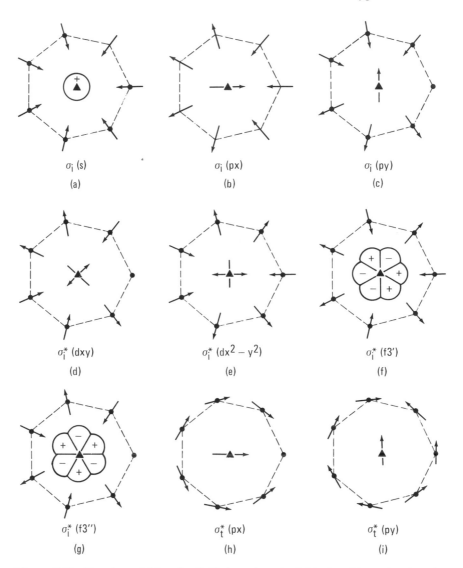

Figure 7-20. Sketches of SOs of $c$-$C_7H_7^+$ as given in Table 7-5. These sketches also represent the SMs of a heptagonal array of atoms. The subscripts i and t refer to "inward" pointing and "tangential" $\sigma$ MOs.

$\sigma_t$(dxy)

(j)

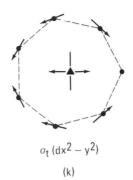

$\sigma_t$ (dx$^2$ − y$^2$)

(k)

$\sigma_t$(f3'')

(l)

$\sigma_t$ (f3')

(m)

$\sigma_t^*$ (j7')

(n)

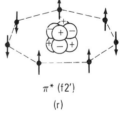

$\pi$(pz)

(o)

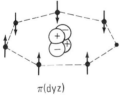

$\pi$(dyz)

(p)

$\pi$(dxz)

(q)

$\pi^*$ (f2')

(r)

$\pi^*$(f2'')

(s)

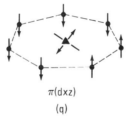

$\pi^*$(g3')

(t)

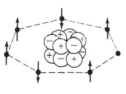

$\pi^*$(g3'')

(u)

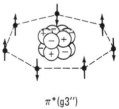

Figure 7-20 (continued)

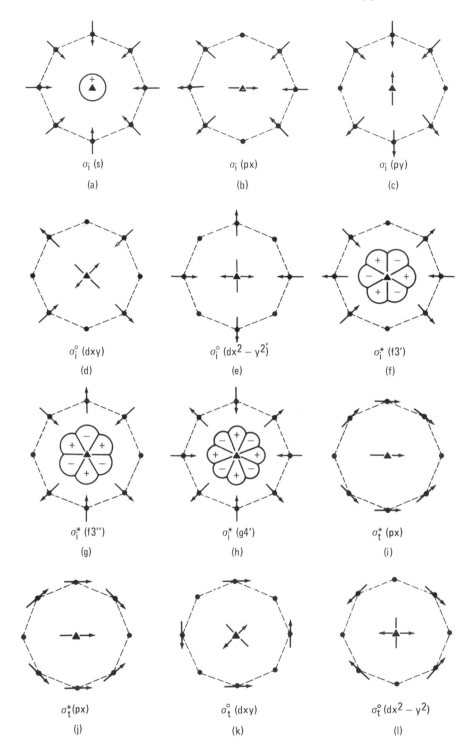

Figure 7-21. Sketches of SOs of $c\text{-}C_8H_8{}^{2-}$ as given in Table 7-6. These sketches also represent the SMs of an octagonal array of atoms. The subscripts i and t refer to "inward" pointing and "tangential" $\sigma$ MOs.

$\sigma_t$ (f3')

(m)

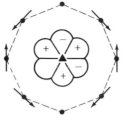

$\sigma_t$ (f3'')

(n)

$\sigma_t$ (g4'')

(o)

$\sigma_t^*$ (k8'')

(p)

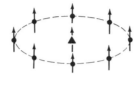

$\pi$(pz)

(q)

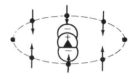

$\pi$(dyz)

(r)

$\pi$(dxz)

(s)

$\pi^\circ$(f2')

(t)

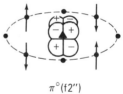

$\pi^\circ$(f2'')

(u)

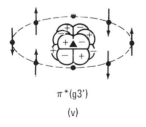

$\pi^*$(g3')

(v)

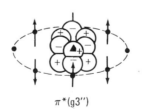

$\pi^*$(g3'')

(w)

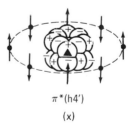

$\pi^*$(h4')

(x)

Figure 7-21 (continued)

Table 7-4. Generator Table for $C_6$ Framework in $c\text{-}C_6H_6$ (18 $e^-$)

| VAO Equivalence Sets | | | | | | | GOs | | | | | | | | |
|---|---|---|---|---|---|---|---|---|---|---|---|---|---|---|---|
| | $s$ | $p_x$ | $p_y$ | $p_z$ | $d_{xy}$ | $d_{x^2-y^2}$ | $d_{xz}$ | $d_{yz}$ | $f2'$ | $f2''$ | $f3'$ | $f3''$ | $g3'$ | $g3''$ | $i6''$ |
| $C_{spi}$ | b | b | b | | a | a | | | | | | a | a | | |
| $C_{px}$ | | a | a | | b | b | | | a | a | b | | | | a |
| $C_{p\pi}$ | | | | b | | | b | b | | | | | | | |

Table 7-5. Generator Table for $C_7$ Framework in $c\text{-}C_7H_7^{+}$ (20 $e^-$)

| VAO Equivalence Sets | | | | | | | GOs | | | | | | | | |
|---|---|---|---|---|---|---|---|---|---|---|---|---|---|---|---|
| | $s$ | $p_x$ | $p_y$ | $p_z$ | $d_{xy}$ | $d_{x^2-y^2}$ | $d_{xz}$ | $d_{yz}$ | $f2'$ | $f2''$ | $f3'$ | $f3''$ | $g3'$ | $g3''$ | $j7''$ |
| $C_{spi}$ | b | b | b | | a | a | | | | | a | | | | |
| $C_{px}$ | | a | a | | b | b | | | a | b | | b | a | | |
| $C_{p\pi}$ | | | | b | | | b | b | a | | | | | a | a |

Table 7-6. Generator Table for $C_8$ Framework in $c\text{-}C_8H_8^{2-}$ (26 $e^-$)

| VAO Equivalence Sets | | | | | | | GOs | | | | | | | | | | |
|---|---|---|---|---|---|---|---|---|---|---|---|---|---|---|---|---|---|---|
| | $s$ | $p_x$ | $p_y$ | $p_z$ | $d_{xy}$ | $d_{x^2-y^2}$ | $d_{xz}$ | $d_{yz}$ | $f2'$ | $f2''$ | $f3'$ | $f3''$ | $g3'$ | $g3''$ | $g4'$ | $g4''$ | $h4''$ | $k8''$ |
| $C_{spi}$ | b | b | b | | n | n | | | a | a | | | a | | a | | | |
| $C_{px}$ | | a | a | | n | n | | | b | b | a | | | a | | b | | |
| $C_{p\pi}$ | | b | | b | | | b | n | n | n | | | | | | | a | a |

Table 7-7.  A Useful Summary of Procedures for Generating Pictorial
Representations of Delocalized and Localized Bonding Views of a Molecule and
for Visualizing its Vibrational Modes

| Delocalized View | Localized View | Normal Modes |
|---|---|---|
| 1. Decide on best electron dash structure using lowest possible formal charges and drawing resonance forms. In such forms, some lp's may be involved in a $\pi$ system. | 1. Decide on which set(s) of doubly occupied delocalized MOs to localize. | 1. Identify the motional atomic vector (AV) equivalence sets and list them in the generator table. |
| 2. Establish the molecular geometry from VSEPR theory if possible. | 2. Hybridize the **GO**s which give rise to the sets(s) in 1. | 2. Using **GO**s, sketch the symmetry motions (SMs). |
| 3. After preassigning lp's *and localized bp's* to (*s*) and suitably oriented (*p*) VAOs, identify all the VAO equivalence sets (hybrids may also be used) and list them in the generator table. | 3. Using the hybridized **GO**s, sketch the SOs. 4. Sketch the MOs, being aware of the consequences of a central atom on which the full set of SOs (hybridized VAOs) is not generated by the hybridized **GO** set. If necessary, rehybridize peripheral atom VAOs *and rotate hybrid* **GO***s* to locate bp's between atoms. | 3. Fill in the rest of the generator table. (i.e., **GO**s and m entries). 4. Take appropriate linear combinations of SMs generated by the same **GO** and sketch all the molecular motions. 5. Identify the normal vibrational modes. |
| 4. Using **GO**s located at a molecular center or centroid, sketch all the SOs, recalling that #SOs = #VAOs. | 5. Draw the resulting MO energy level diagram and occupy the appropriate levels with electrons. | |
| 5. Fill in rest of generator table (i.e., the necessary **GO**s and the *n, a,* or *b* entries). Establish partner **GO** and SO sets. | 6. Verify that the bond order per link is the same as in the delocalized view. | |
| 6. Take linear combinations of SOs generated by the same **GO** and sketch all the MOs. | | |
| 7. Draw an MO energy level diagram and occupy the appropriate levels with electrons. | | |
| 8. Deduce the bond order per link. | | |

PROBLEMS

1. Draw the localized MO energy level diagrams for (a) $Te_4^{2+}$, (b) rectangular $c$-$C_4H_4$, and (c) $c$-$C_5H_5^-$.

2. The cyclic molecule $N_2S_2$ has been shown to have a planar diamond-shaped structure. (a) On the basis of electron dash structures, would you expect this molecule to have an N—N and an S—S link, or alternating S—N links? (b) Develop the delocalized view of this molecule.

3. One of the singlet (i.e., diamagnetic) state structures of cyclobutadiene is square planar. Develop the delocalized view of this molecule. Comment on the relationship of the electrons in the degenerate nonbonding levels.

4. Develop the delocalized view of the cyclic planar molecule $(NS)_3^-$ which possesses alternating N—S links.

5. Sketch the molecular vibrations of square planar cyclobutadiene and give the corresponding GO table.

6. Sketch the delocalized MO energy level diagrams for (a) $c$-$C_6H_6$, (b) $c$-$C_7H_7^+$, and (c) $c$-$C_8H_8^{2-}$.

7. Give the localized views of the species in Problem 6.

8. Develop the delocalized and localized views of $Te_4^{2+}$ using an axis system in which the $x$ and $y$ axes point at the atoms. Compare the results with those obtained in the text.

# CHAPTER 8

# Octahedral and Related Molecules

Octahedral species include not only examples such as $PF_6^-$ and $CoF_6^{3-}$ which contain central atoms but also those which do not, such as the cluster systems $B_6H_6^{2-}$ and $Mo_6Cl_8^{4+}$. We will also find it instructive to consider in this Chapter $BrF_5$, $ICl_4^-$, and $B_5H_9$ whose structures, depicted in Figure 8-1, can be considered as missing one or two octahedral vertex atoms.

## 8.1. $ICl_4^-$

Rather than picking a molecule with octahedral symmetry for our first example, we will find it worthwhile to begin with the square planar $ICl_4^-$ ion. The square planarity of this ion is easily verified by examining its electron dash structure in Figure 8-2. The two lp's on iodine repel one another more strongly than lp's repel bp's and so the lp's are 180° apart while the bp's are separated by 90°. Why do we place these lp's on iodine rather than distribute them among the chlorines? Iodine is a larger atom and it is thus better able to accommodate electron pair repulsions than chlorine.

To obtain the delocalized view of $ICl_4^-$, we first assign the chlorine lp's to (Cl3$s$), (Cl3$px$), and (Cl3$pz$) VAOs and the iodine lp's to (I5$s$) and (I5$pz$) VAOs. The remaining chlorine equivalence set Cl$pr$, by analogy to that in $Te_4^{2+}$, leads to the four SOs listed in Generator Table 8-1. The eight electrons are easily seen to reside in a pair of degenerate 5-center 2-electron BMOs generated by $px$ and $py$ GOs, and in a pair of nondegenerate NBMOs generated by $s$ and $dxy$ GOs. Note that as with $Te_4^{2+}$, our axis system for $ICl_4^-$ (Figure 7-2) points the $x$ and $y$ GO axes *between* the peripheral atoms to maximize delocalization. It is left as an exercise to show that directing these

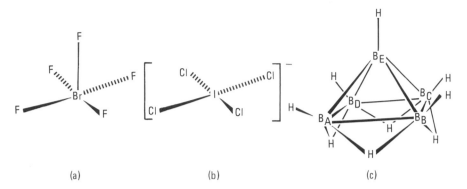

Figure 8-1.  Structures of (a) $BrF_5$, (b) $ICl_4^-$, and (c) $B_5H_9$.

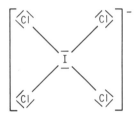

Figure 8-2.  Electron dash structure for $ICl_4^-$.

Table 8-1.  Generator Table for $ICl_4^-$ without
Iodine (d) VAOs (8 e$^-$)

|                       |     | GOs |     |     |
|-----------------------|-----|-----|-----|-----|
| VAO Equivalence Sets  | s   | px  | py  | dxy |
| I$pr$                 |     | n   | n   |     |
| Cl$pr$                | n   | n   | n   | n   |

axes toward two of the iodines gives rise to a 3-center rather than a 5-center bond system.

Localization of the four occupied delocalized MOs requires that an $sp^2d$ GO set be employed. A computer-generated contour diagram of a member of such a set is shown in Figure 8-3. The major lobe of each member of such a set points to the corner of a square. This implies that we can orient the set so that the major lobes point to the chlorines. A member of this GO set is schematically represented in Figure 8-4a as is the SO it generates in the Cl$pr$ equivalence set. Since the average bond order in the delocalized BMOs is 0.5, an antibonding contribution to each of the four localized MOs is present. It is seen in Figure 8-4b that this antibonding contribution comes from the (Cl3$py$) VAO diagonally opposite to the (Cl3$py$) VAO making the major contribution to the localized bond. As a consequence of pointing the GOs toward the

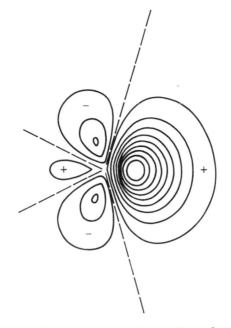

Figure 8-3. Computer-generated plot of an $sp^2d$ orbital.

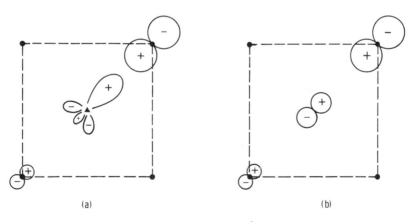

(a)                                   (b)

Figure 8-4. Chlorine ($3py$) VAOs called in by an $sp^2d$ **GO** (a) and the ($5p$) VAO such a **GO** calls in on the iodine (b).

chlorine atoms, the ($5p$) VAOs of the iodine have merely been rotated by 45° from the $x$ and $y$ axes. Because of the placement of the node in the **GO** in Figure 8-4a, the remaining two (Cl3$py$) VAOs adjacent to the major (Cl3$py$) VAO are not called into the localized MO. The iodine lone pairs can be localized in a pair of digonal ($spz$) VAO hybrids.

Many molecules exist in which an atom is surrounded by more than the

Table 8-2. Generator Table for $ICl_4^-$ including Iodine (d) VAOs (8 $e^-$)

| | GOs | | | |
|---|---|---|---|---|
| VAO Equivalence Sets | $dz^2$ | $px$ | $py$ | $dx^2 - y^2$ |
| $I$pr = ($I5px$), ($I5py$) | | n | n | |
| $Idz^2$ = ($I5dz^2$) | n | | | |
| $Idx^2 - y^2$ = ($I5dx^2 - y^2$) | | | | n |
| Cl*pr* = ($Cl_A3py$), ($Cl_B3py$), ($Cl_C3py$), ($Cl_D3py$) | n | n | n | n |

usual octet of electrons in sigma-type MOs plus any lp's that may be present. Such atoms are said to be *hypervalent*.

For hypervalent atoms, it is, of course, always possible to invoke the use of (*d*) VAOs for the purpose of making localized 2-center 2-electron bonds. In the present example containing hypervalent iodine, the only iodine (*d*) VAOs possessing lobes which directly face the (*py*) VAOs of the chlorines are ($I5dx^2 - y^2$) and the ($I5dz^2$). (In the latter orbital, the "doughnut" lies in the plane of the chlorine (*py*) VAOs.) Since ($I5dz^2$) is generated by a $dz^2$ **GO**, we have introduced this **GO** in our Generator Table 8-2 and for similar reasons we have also introduced $dx^2 - y^2$ in this table. The latter **GO** poses no particular problem, but we see immediately that $dz^2$ also generates an SO in Cl*pr*. Since four SOs have already been generated in that set (see Table 8-1) we know that the SO $\tilde{\sigma}(dz^2|\text{Cl}pr)$ cannot be independent and, indeed, it is easily shown that this SO is identical to ($s$|Cl*pr*) in Table 8-1. As we have seen previously, it is appropriate to label the corresponding column of the **GO** table by the most inclusive **GO** (in this case $dz^2$). From Table 8-2 we see that a BMO and an ABMO can be obtained by linear combination of the SOs generated by the same **GO**. The four delocalized BMOs which result are occupied by the eight electrons available for bonding.

For the corresponding localized view we note that the hybrid **GO** set associated with the occupied delocalized BMOs is $p^2d^2$. As might be expected, the contours of a member of such a set look qualitatively very much like those of the $sp^2d$ hybrid orbital shown in Figure 8-3. This **GO** hybrid set generates the corresponding ($p^2d^2$) iodine VAO set, whose main lobes point toward the chlorines. Four localized BMOs can be formed from these iodine ($p^2d^2$) hybrids and the (*py*) VAOs of the chlorines.

The molecular motions of square planar $ZY_4$ species are obtained by considering the SMs of the three AVs on each member of the $Y_4$ moiety (see $Te_4^{2+}$) and the three AVs on Z. The visualization of these motions is left as an exercise at the end of the chapter.

## 8.2. $BrF_5$

From the electron dash structure in Figure 8-5 we conclude via the VSEPR rules that this molecule is tetragonal pyramidal. In contrast to $ICl_4^-$, in which two atoms are missing from octahedral vertices, $BrF_5$ has only one such atom

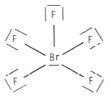

Figure 8-5. Electron dash structure of BrF$_5$.

Table 8-3. Generator Table for BrF$_5$ (10 e⁻)

| VAO Equivalence Sets | GOs | | | |
|---|---|---|---|---|
| | $s$ | $px$ | $py$ | $dxy$ |
| Br$pa$ = (Br4$pz$) | n | | | |
| Br$pb$ = (Br4$px$), (Br4$py$) | | n | n | |
| F$pa$ = (F$_E$2$pz$) | n | | | |
| F$pb$ = (F$_A$2$pz$), (F$_B$2$pz$), (F$_C$2$pz$), (F$_D$2$pz$) | n | n | n | n |

missing. Other examples of such structures are XF$_5$ (X = Cl, I), and ChF$_5^-$ (Ch = S, Se) in which the chalcogen, like the bromine in BrF$_5$, is also hypervalent. In the molecular structure of BrF$_5$, the basal fluorines are bent out of the plane toward the apical fluorine, and their angle with the apical fluorine is 84.4°. This distortion can be rationalized as stemming from lp–bp repulsions which exceed bp–bp repulsions. Another interesting feature of this molecule is the difference in length between the axial bond (1.744 Å) and the equatorial bonds (1.689 Å). It is apparent from the structure of BrF$_5$ that the apical fluorine experiences a molecular potential different from that experienced by the four equivalent basal fluorines. Thus we must treat the two types of fluorines separately in generating our SOs. Because of fluorine's greater electronegativity, there is some ionic character in the Br–F bonds.

We begin developing our delocalized view by assigning all the fluorine lp's to (2$s$), (2$px$), and (2$py$) VAOs and the bromine lp to the (4$s$) VAO. We can construct Generator Table 8-3 wherein the labels "a" and "b" in the VAO equivalence sets denote the apical and basal positions. For simplicity we will assume the Br and basal fluorine atoms to be coplanar. The GO lies somewhere along the Br–F$_{ax}$ internuclear axis. The main contributor to the NBMO generated by $s$ stems from the $\tilde\sigma(s|$F$pb)$ SO. That this is so can be visualized by noting that the two remaining SOs strongly interact (owing to their colinear configuration) and they linearly combine to give a BMO and an ABMO. On the other hand, a linear combination of two orbitals which includes $\tilde\sigma(s|$F$pb)$ as one of the contributors leads to relatively little interaction owing to very poor overlap with either $\tilde\sigma(s|$Br$pa)$ or $\tilde\sigma(s|$F$pa)$.

The ten electrons fill the three BMOs and two NBMOs, giving an average bond order of 1/2 in each link in the basal plane and a bond order of 1.0 in the apical link. This conclusion is at variance with the rule (usually valid) that strong bonds are associated with short bond distances, because the apical bond distance is slightly longer than the equatorial bond distance. This anomaly presumably originates in nonbonding repulsions arising from the proximity of the basal to the apical fluorines, which in turn results from the fact that the basal fluorines are actually bent out of the plane toward the apical fluorine due to the predominance of lp–lp bond pair repulsions over bp–bp repulsions. In order for the bromine lone pair to exert this effect, it must possess some $(4pz)$ character which gives it directionality away from the apical fluorine. Correspondingly, the VAO in the VAO equivalence set, Br$pa$, has some $(4s)$ character.

Since $\sigma(s)$ is already a localized occupied 2-center MO, we confine our attention to the remaining occupied MOs which bond in the basal plane. These are associated with the **GO** set $s, px, py$, and $dxy$ which when hybridized produce four equivalent hybrid **GO**s in a square planar arrangement. Since the Br$pb$ VAO equivalence set contains only two orbitals, the four hybrid **GO**s cannot call in orthogonal contributions from the bromine atom, and consequently the localized MOs have substantial antibonding character, reducing the bond order in each basal Br—F link to 1/2.

A discussion of the molecular motions of square pyramidal ZY$_5$ molecules such as BrF$_5$ is deferred until the end of the next section.

## 8.3. B$_5$H$_9$

Most neutral boron hydrides such as B$_5$H$_9$ can be obtained by heating the simplest member, gaseous B$_2$H$_6$, and eliminating hydrogen gas. In the present discussion we set out to obtain an understanding of the bonding in B$_5$H$_9$. The structure of this molecule in Figure 8-1c shows that two kinds of hydrogens are present, namely, those which are bound to one boron and those which are linked to two boron atoms. The hydrogens of the former type, of which there are five, are called *terminal hydrogens* and those of the latter variety, of which there are four, are referred to as *bridging hydrogens*. Because all the hydrogens are not of the terminal type, as in $c$-C$_5$H$_5^-$, we must first appreciate how the bonding in a bridging B—H—B bond differs from that in a terminal B—H bond before determining which boron VAOs to consider and how many electrons are available to bind the square pyramidal cluster of borons together. As we will see later in this section, there is an insufficient number of electrons to bind each adjacent boron atom with two electrons. Thus a normal Lewis structure cannot be drawn. Models for predicting boron hydride structures have been put forth but they are still in a developmental stage.

Bridging hydrogens can be linear, as in FHF$^-$, or bent, as in many boron and transition metal hydrides. Two bent bridging hydrogens occur in the simplest boron hydride B$_2$H$_6$, a gas under ordinary conditions, which can be made from BCl$_3$ and LiAlH$_4$. The structure of this molecule, depicted in Figure 8-6, consists of two boron atoms which are surrounded by four hydrogens in an approximately tetrahedral fashion. The nearly tetrahedral stereochemistry around the borons makes it convenient to utilize on each boron two

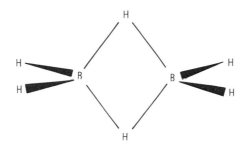

Figure 8-6. Structure of $B_2H_6$. The molecule has four terminal and two bridging hydrogens.

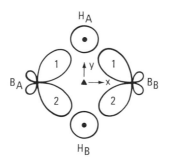

Figure 8-7. Boron and bridging hydrogen VAOs available for the MO system of the $B_2H_2$ ring of $B_2H_6$.

Table 8-4. Generator Table for $B_2H_2$ Ring in $B_2H_6$ (4 e$^-$)

|  | GOs | | | |
|---|---|---|---|---|
| VAO Equivalence Sets | $s$ | $px$ | $py$ | $dxy$ |
| $Bsp$-in $= (B_A sp^\beta 1), (B_A sp^\beta 2), (B_B sp^\beta 1), (B_B sp^\beta 2)$ | n | n | n | n |
| $H1s = (H_A 1s), (H_B 1s)$ | n |  | n |  |

($Bsp^x$) VAOs which are directed at the ($1s$) VAO of each of the four terminal hydrogens. The four 2-center BMOs which can be formed by linear combination of each pair of neighboring ($H1s$) and ($Bsp^x$) VAOs will be reserved for four pairs of bonding electrons holding the terminal hydrogens to the boron. The remaining four valence electrons of the total of twelve available in this molecule are used to bind the borons together via the two bridging hydrogens. The two remaining VAOs on each boron are completely determined by ortho gonality if we choose them to be equivalent. From Figure 8-7 it is seen that they are approximately directed toward the bridging hydrogens. Using Figure 8-7 we construct Generator Table 8-4. From this table we see that linear combinations of the SOs generated by $s$ and $py$ each give a bonding and an antibonding MO. The corresponding MO energy diagram in Figure 8-8 reveals that both 4-center BMOs are filled with the four valence electrons we have available, giving an average bond order of 0.5 per B—H link in the $B_2H_2$ ring

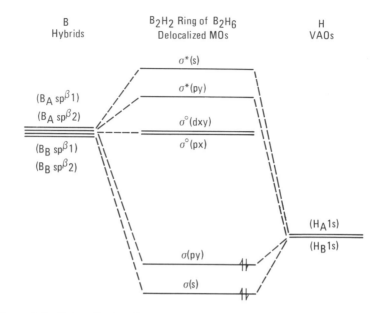

Figure 8-8. Delocalized MO energy level diagram for the $B_2H_2$ ring of $B_2H_6$.

system. It should be noted that the delocalized bonding picture of these bridging hydrogens differs from that in $FHF^-$ wherein the NBMO is also occupied. In the case at hand, the NBMO associated with each bridging hydrogen is empty.

To obtain a localized picture of the bonding in the $B_2H_2$ ring of $B_2H_6$, we proceed in the usual manner by hybridizing the **GO**s associated with the occupied MOs in Figure 8-8. The resultant digonal *sp* hybrid **GO**s have their main lobes directed at $H_A$ and $H_B$, which informs us that one bonding pair of electrons is localized near $H_A$ and the other link near $H_B$. These bond pairs must each link both borons and one bridging hydrogen in a 3-center 2-electron bond formed by the overlap of two boron ($sp^\beta$) hybrid VAOs and a hydrogen (1$s$) valence orbital. Note that we do not choose here to point the lobes of the *sp* **GO** set between diagonally located B—H links. Such a choice is, of course available and this is left as an exercise. As expected, the average bond order is 0.5 per B—H link. Again, note the difference in the localized bond pictures of $FHF^-$ and $B_2H_6$. In $FHF^-$, *two* localized pairs of electrons give rise to an overall bond order of 0.5 per H—F link. This is because of the partial antibonding contribution in each localized 3-center bond.

One of the reasons boron hydrides are typically reactive to Lewis bases is that the NBMOs in the B—H—B linkages are empty and hence receptive to attack by electron pair donors. Thus $B_2H_6$ is cleaved by two $(CH_3)_3N$ molecules to give two molecules of $(CH_3)_3NBH_3$ in which no NBMOs are present.

Three-center bonds containing a bridging hydrogen between two borons always contain a pair of electrons in a delocalized BMO of the type just described. With this generality it becomes possible to assign a pair of electrons to each such 3-center system we encounter in a new molecule. We now proceed to do this in our treatment of $B_5H_9$. When B—H—B linkages occur in boron hydride systems, a complete description of the bonding in the localized view always leads to 3-center 2-electron MOs of the type just discussed. Anticipating this we can reserve the hydrogen $(1s)$ VAOs and the appropriate $(sp^\beta)$ hybrid VAO on the borons to form such bonds when they are present. Thus for $B_5H_9$ we reserve two $(Bsp^\beta)$ VAOs on each basal boron for the formation of the four localized B—H—B bonds. These B—H—B bonds will require a total of eight of the twenty-four available valence electrons.

For the apical atom, $B_E$, if we reserve the upwardly directed $(sp^{\alpha'})$ hybrid AO for bonding to the apical hydrogen, then the inwardly directed, orthogonal hybrid $(sp^{\beta'})$ is available for bonding the boron cluster and it is the sole member of the VAO equivalence set Bspai. The AOs $(B_E 2px)$ and $(B_E 2py)$ form the VAO equivalence set Bpt. For each basal boron atom we reserve an $(sp^\alpha)$ hybrid orbital for bonding to the terminal hydrogen atom and two $(sp^\beta)$ hybrid orbitals for contributions to two 3-centered B—H—B bonds. There remains a linear combination of the $(2s)$ and $(2p)$ boron orbitals which is orthogonal to the reserved orbitals and is available for bonding the boron cluster. These inwardly directed hybrid orbitals, which we call $sp^\gamma$, belong to the VAO equivalence set denoted by Bspbi. Using the axis system in Figure 8-9 in which the **GO** center lies above the basal boron plane, Generator Table 8-5 is easily constructed. The SO $\tilde{\sigma}(dxy|Bspi)$ is identical to the antibonding

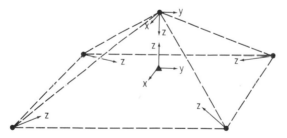

Figure 8-9. Axis system for the square pyramidal $B_5$ cluster of $B_5H_9$.

Table 8-5. Generator Table for $B_5$ Cluster in $B_5H_9$ (6 e$^-$)

| VAO Equivalence Sets | GOs | | | |
| --- | --- | --- | --- | --- |
| | $s$ | $px$ | $py$ | $dxy$ |
| $Bspai = (B_E sp^{\beta'})$ | n | | | |
| $Bpt = (B_E 2py)(B_E 2px)$ | | n | n | |
| $Bspbi = (B_A sp^\gamma), (B_B sp^\gamma), (B_C sp^\gamma), (B_D sp^\gamma)$ | b | n | n | a |

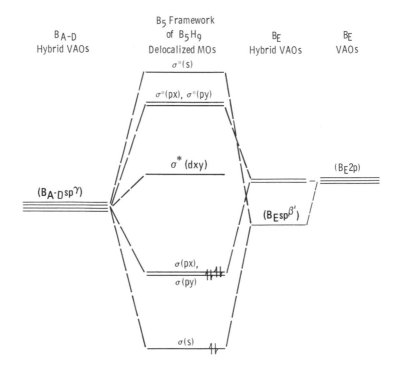

Figure 8-10. Delocalized MO energy level diagram for the $B_5$ framework of $B_5H_9$.

MO $\sigma^*(dxy)$ and by linear combination of the remaining SOs we form three BMO–ABMO pairs. Two of these pairs are degenerate since they arise from $px$ and $py$ **GO** partners. The five localized BH MOs and the four BHB MOs together require eighteen electrons. The remaining six valence electrons fill the three 5-center delocalized BMOs, as shown in the delocalized energy level diagram for the $B_5$ cluster in Figure 8-10. The overall bond order is 3/8 since there are eight B—B links in the a square pyramidal boron cluster.

The three bonding pairs in Figure 8-10 are localized by means of an $sp^2$ **GO** hybrid set. The resulting localized MOs are multicentered because of the unusually low bond order of the boron cluster. This molecule serves as yet another example of our observation that as the localized orbitals become increasingly multicentered they generally become more difficult to visualize. This "delocalization" renders the localized view less useful as a complement to the delocalized view.

The molecular motion of square and tetragonal pyramidal molecules will now be briefly discussed. The motions of $ZY_5$, of which $BrF_5$ is an example, are obtained by taking linear combinations of the SMs of the basal Y atoms, the apical Y atom, and the Z atom that are generated by the same **GO**. The motions of square pyramidal $Z_5$ are similarly obtained. Since the motions of the component moieties of such molecules have already been examined, the generation of their LCAVs will be left as an exercise.

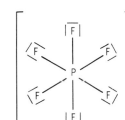

Figure 8-11. Electron dash structure of PF$_6^-$.

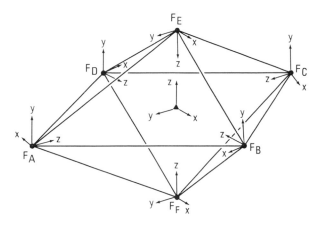

Figure 8-12. Axis system for PF$_6^-$.

## 8.4. PF$_6^-$

The electron dash structure of PF$_6^-$ in Figure 8-11 tells us that there are no lp's on phosphorus and that this hypervalent atom has twelve electrons surrounding it. VSEPR theory predicts that these six equivalent electron pairs will direct themselves toward the apices of a regular octahedron. A planar hexagonal array of fluorines around a phosphorus would be much less favorable because such a geometry dictates only 60° between electron pairs. Owing to the greater electronegativity of fluorine, there is considerable ionic character in the P—F bonds.

We first assign the lone pairs to the $(2s)$, $(2px)$, and $(2py)$ VAOs on each fluorine. Using the axis system in Figure 8-12 to construct Generator Table 8-6, we see that four of the six electron pairs available for bonding are housed in four BMOs arising from the linear combinations of four pairs of SOs generated by the same **GO**. The remaining two electron pairs reside in the nonbonding SOs (MOs) $\tilde{\sigma}(dx^2 - y^2)|Fpr)$ and $\tilde{\sigma}(dz^2|Fpr)$. The overall bond order is 2/3.

Contrary to our usual procedure, all of the **GO** axes in Figure 8-12 are

Table 8-6. Generator Table for $PF_6^-$ ($12\,e^-$)

| VAO Equivalence Sets | GOs | | | | | |
|---|---|---|---|---|---|---|
| | $s$ | $px$ | $py$ | $pz$ | $dx^2 - y^2$ | $dz^2$ |
| $Ps = (P3s)$ | n | | | | | |
| $Ppr = (P3px), (3py), (P3pz)$ | | n | n | n | | |
| $Fpr = (F_A2pz), (F_B2pz), (F_C2pz), (F_D2pz),$ | | | | | | |
| $\qquad (F_E2pz), (F_F2pz)$ | n | n | n | n | n | n |

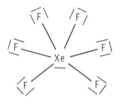

Figure 8-13. Electron dash structure for $XeF_6$.

pointed toward (rather than between) atoms. Although geometrically we are unable to direct all three Cartesian axes between atoms, we could have pointed two of them in such directions (e.g., $x$ and $y$). Three reasons why the axis system shown is more convenient are: 1) the equivalence of the **GO** Cartesian axes in the octahedron is more apparent, 2) the SOs among tangential orbitals (e.g., those among the $Fpt$ set in $CoF_6^{3-}$ to be discussed shortly) are easier to visualize, and 3) the two partner sets among the (d) VAOs on a central atom are more readily recognized (see later).

Inspection of Table 8-6 reveals that the hybrid **GO** set for localizing the occupied delocalized MOs is $sp^3d^2$. The members of such a set point to the apices of an octahedron. Thus we wish to construct a localized MO in each P—F link. The only SOs on phosphorus which can be called in by the $sp^3d^2$ **GO** hybrid set are in the $Ps$ and $Ppr$ equivalence sets. Thus, each main lobe of an $sp^3d^2$ hybrid **GO** can call in one member of a phosphorus digonal $sp$ hybrid directed toward a fluorine. Each of these digonal hybrids is constructed from the (P3s) and one of the (P3p) VAOs, leading to the linear combinations $a(P3s) \pm b(P3px)$, $a(P3s) \pm b(P3py)$, and $a(P3s) \pm b(P3pz)$ ($b/a \approx 3^{1/2}$). Because full participation of the VAOs on phosphorus is not possible, these linear combinations are not orthogonal. To assure orthogonality of the localized MOs, therefore, each MO is bonding in one link and antibonding in the opposite link. The result is an average bond order of $2/3$ per P—F link.

Species with forty-eight valence electrons such as $PF_6^-$ are expected to be octahedral on the basis of VSEPR theory. There currently does not exist a simple way (such as is provided by VSEPR theory) to predict the structure of fifty-electron hypervalent systems such as $SbBr_6^{3-}$, $SeBr_6^{2-}$, and $XeF_6$. The electron dash structure for $XeF_6$ in Figure 8-13 involves seven electron pairs around the xenon atom. Experimentally it is found in several systems, such as $TeCl_6^{2-}$ and $SbCl_6^{3-}$, that the structures are essentially octahedral with little

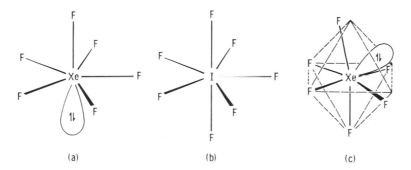

Figure 8-14. Possible structure for $XeF_6$ (a). Experimentally observed structures for $IF_7$ (b) and $XeF_6$ (c).

if any significant distortion. By contrast, other fifty-electron systems (such as $XeF_6$ in the gas phase) display structural and spectroscopic properties which are consistent with some distortion. Because the distorted octahedral systems possess very low symmetry, we will not consider their bonding patterns.

The bonding in fifty-electron molecules which are truly octahedral can be rationalized in the delocalized view by allowing the two additional electrons to occupy an ABMO. This occupied ABMO must be $s$-generated since only $s$ of all the GOs contains the full symmetry of the octahedron. If these two electrons were placed in a triply degenerate $p$-generated ABMO set, only two of the MOs could be occupied by one electron each and the full symmetry of the octahedron could not be contained in two half-occupied MOs because of the lack of occupation of the third member of the partner set. Not only would the molecule be paramagnetic (in contradiction to experiment) but it would also tend to undergo Jahn–Teller distortion away from octahedral symmetry.

It would seem logical for $XeF_6$ to have the pentagonal pyramidal structure depicted in Figure 8-14a because such a structure is consistent with the experimentally observed structure of $IF_7$, wherein the central atom is surrounded by seven bond pairs as shown in Figure 8-14b. However, the best information we presently have indicates that the structure of $XeF_6$ in the gas phase resembles the distorted octahedron shown in Figure 8-14c, in which the lone pair apparently protrudes through a triangular face. These results point up some of the limitations of VSEPR concepts and suggest the need for more theoretical work on systems with seven electron pairs around a central atom.

The molecular motions of octahedral $ZY_6$ species will be discussed at the end of this chapter.

## 8.5. Octahedral Transition Metal Complexes— Some Special Considerations

Transition metal complexes are capable of using their incompletely filled ($d$) orbitals in the $n - 1$ shell for interaction with *ligand* orbitals. Here a *ligand* is an atom or molecule which binds to the metals and $n$ refers to the principal quantum number of the valence shell. Thus, for example, in addition to the (4s) and (4p) valence orbitals of the first-row transition elements, scandium through copper, we also include the (3d) orbitals as valence orbitals because they are of similar energy. The same relationship holds for the (d) orbitals in the penultimate shell and the AOs in the outermost shell in the two succeeding rows of the periodic chart. The lanthanides (cerium through lutetium) and actinides (thorium through lawrencium) of the inner transition series are characterized by the filling of the (4f) and (5f) shells, respectively. The extent of (f) orbital participation in bonding, however, is presently not as well understood as that of (d) VAOs in the transition elements.

Lewis and others initiated the idea that metal atoms could accept lone pairs from ligands such as $Cl^-$ and $H_2O$ to form a *coordinate covalent link*. The terms *"coordination compounds"* or *"coordination complexes"* are used to describe such species. A localized bonding approach which stemmed from these ideas was popularized by L. Pauling during the 1930s and 1940s. However, the delocalized view of many coordination compounds is best visualized through the *ligand field theory* (LFT) which was developed by several theoretical physicists and chemists between 1930 and 1950. LFT is a refinement of an electrostatic theory called *"crystal field theory"* (CFT) which treats the ligands as negative point charges or point dipoles surrounding a positively charged metal atom. The crystal field model, then, considers the metal–ligand interactions as strictly ionic. It is, of course, certainly more realistic to include, at least to some degree, covalent interaction which involves electron sharing via orbital overlaps and MO formation. This approach is fundamental to the LFT which we will discuss after CFT. In further defense of LFT it should be noted that coordinated atoms are not points, as is assumed in CFT, since their sizes are quite comparable to the metal atom to which they are attached. It should also be pointed out that VSEPR considerations are not very useful in discussing structures of transition metal complexes.

The instructiveness of CFT is demonstrated by briefly considering an octahedral complex such as $TiF_6^{3-}$. Anions of this type are generally formed by treating the neutral metal halide with excess halide ion. In $TiF_6^{3-}$ the trivalent metal is surrounded by six fluoride ions in an octahedral array. In the *free* gaseous $Ti^{3+}$ ion (i.e., no ligands present) the single valence electron can occupy any of the five (3d) VAOs with equal probability. The same would be true if six negative ligand charges were "smeared out" in a sphere around the metal, except that all the (d) VAOs would be equally raised in energy owing to repulsions between the ligand charges and the (d) electron on the metal. If these six negative ligand charges arrange themselves into an octahedral array, however, they will tend to repel the electron density only in those (d) orbitals whose lobes are directed toward the charges [e.g., the $(dx^2 - y^2)$ and $(dz^2)$ VAOs in the axis system shown in Figure 8-12. Less repulsion will occur with orbital lobes directed between the charged ligands [i.e., the $(dxy)$, $(dxz)$, and $(dyz)$ VAOs]. It can be shown that the rearrangement of charges from spherical symmetry to octahedral symmetry does not change the overall energy of

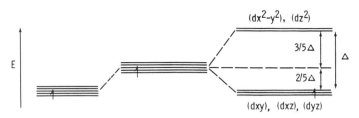

Figure 8-15. The relative energies of the $(d)$ VAOs of a gaseous $Ti^{3+}$ ion which is (a) free, (b) surrounded by six ligand charges in a spherical array, and (c) surrounded by six ligand charges in an octahedral geometry.

the $(d)$ VAOs. Consequently, two of the $(d)$ orbitals rise in energy by $(3/5)\Delta$ each while three descend by $(2/5)\Delta$ each. These orbital energy changes are depicted in Figure 8-15. That the $(3dxy)$, $(3dxz)$, and $(3dyz)$ VAOs are partners in an octahedral field is obvious from the facts that they are all equivalent in octahedral symmetry, and that none of them by itself possesses the full symmetry of the octahedron. Since neither the $(3dz^2)$ nor the $(3dx^2 - y^2)$ orbitals have the full symmetry of the octahedron, they must form a second degenerate partner set. The titanium $(d)$ VAO electron resides in the three degenerate orbitals since they lie at lower energy than the doubly degenerate set. Excitation of this electron by ultraviolet light of energy $\Delta$ promotes the electron to the higher-energy degenerate pair of orbitals wherein the electron encounters more severe repulsions with the ligand charges.

An interesting consequence of the splitting of the $d$ VAOs is that an octahedral complex with four $(d)$ VAO electrons, for instance, may contain either four or two unpaired $(d)$ VAO electrons, depending on whether or not $\Delta$ is smaller or larger than the electron pairing energy. Coordination complexes in which spin pairing occurs in the lower $(d)$ orbitals are called "strong field" (i.e., $\Delta$ is large relative to the spin pairing energy) while those in which electrons remain unpaired by occupying the upper set are referred to as "weak field."

The need for covalency in coordination compounds can be appreciated by realizing that it is unrealistic to consider the existence of a "naked" $Mo^{5+}$ ion in $MoCl_5$, for example, because the complete transfer of five electrons from a molybdenum atom to the chlorine ligands would result in an intolerably large positive charge on the metal. Substantial electron sharing between the chlorine atoms and the metal is thus expected, in order to reduce the charge.

There are many results from a variety of experiments which strongly demonstrate that electrons associated with the metal atom or ion in the uncomplexed state are partially delocalized onto ligand atoms in the complex and that ligand electron density is simultaneously shifted to the metal. In LFT this sharing of electron density is accounted for by MO formation between ligand and metal atom VAOs. We have now returned to familiar ground, because this procedure is just the one which we have been following for a variety of geometries containing a central atom, including the octahedron.

The major difference is that we now *must include penultimate (d) orbitals in MO formations* because in transition metals these AOs are energetically compatible with the valence $(s)$ and $(p)$ VAO set.

## 8.6. $CoF_6^{3-}$ and $Co[P(OCH_3)_3]_6^{3+}$

We discuss these two transition metal complexes in some detail because, as we shall see, they display rather contrasting electronic properties despite the fact that both are octahedral and contain cobalt in the same oxidation state $(3+)$. Formally the $CoF_6^{3-}$ ion can be thought of as being made up of a $Co^{3+}$ ion and six fluoride ions. Alternatively, it can be conceived of as possessing a cobalt atom, six fluorine atoms, and three negative charges. A Lewis structure can be drawn, as shown in Figure 8-16, which emphasizes the covalent character of this ion. Note that VSEPR theory is not helpful in predicting the structure of $CoF_6^{3-}$. Recall that in the crystal field approach all of the bp's become fluoride lp's, and the cobalt lp and lone singles reside in $(3d)$ VAOs on the $Co^{3+}$ ion. This approach stresses the contribution of the ionic interactions between the cobalt metal atom and the electronegative fluorines. The covalent electron dash structure in Figure 8-16 tells us that the formal charge on each fluorine is zero. The formal charge on cobalt, which formally possesses twelve electrons, is $-3$, however, since a neutral cobalt atom has only nine electrons. The $-3$ formal charge accounts for the overall charge on the ion.

As we will see shortly, the inclusion of $(3d)$ orbitals in the cobalt VAO set permits $\pi$ interactions to occur between some of these VAOs and the tangential $(2p)$ VAOs of the fluorines. Therefore, we reserve only the $(F2s)$ VAOs for a lone pair on each of the ligands. Although it is tempting to reserve the $(4s)$ VAO on cobalt for a lone pair, we refrain from doing so because the $(4s)$ VAO is not the lowest-energy VAO on the metal. Thus we must consider the possibility of using a $(3d)$ VAO for this purpose and we will do this later.

In the axis system shown in Figure 8-12, the **GO** axes are directed toward the ligand atom centers. This arrangement is advantageous because it is

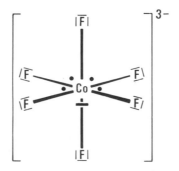

Figure 8-16. Lewis structure for $CoF_6^{3-}$.

Table 8-7. Generator Table for $CoF_6{}^{3-}$ (42 e⁻)

| VAO Equivalence Sets | GOs | | | | | | | | | | | | | | |
|---|---|---|---|---|---|---|---|---|---|---|---|---|---|---|---|
| | $s$ | $px$ | $py$ | $pz$ | $dx^2-y^2$ | $dz^2$ | $dxy$ | $dxz$ | $dyz$ | $f1'$ | $f1''$ | $f2'$ | $g1'$ | $g1''$ | $g4''$ |
| Codr = (Co3dx²−y²), (Co3dz²) | | | | | n | n | | | | | | | | | |
| Coda = (Co3dxy), (Co3dxz), (Co3dyz) | | | | | | | n | n | n | | | | | | |
| Cos = (Co4s) | n | | | | | | | | | | | | | | |
| Copr = (Co4px), (Co4py), (Co4pz) | | n | n | n | | | | | | | | | | | |
| Fpr = (F_A2pz), (F_B2pz), (F_C2pz),(F_D2pz), (F_E2pz), (F_F2pz) | | n | n | n | n | n | | | | | | | | | |
| Fpt = (F_A2px), (F_A2py), (F_B2px), (F_B2py), (F_C2px), (F_C2py), (F_D2px), (F_D2py), (F_E2px), (F_E2py), (F_F2px), (F_F2py) | n | n | n | n | | | n | n | n | n | n | n | n | n | n |

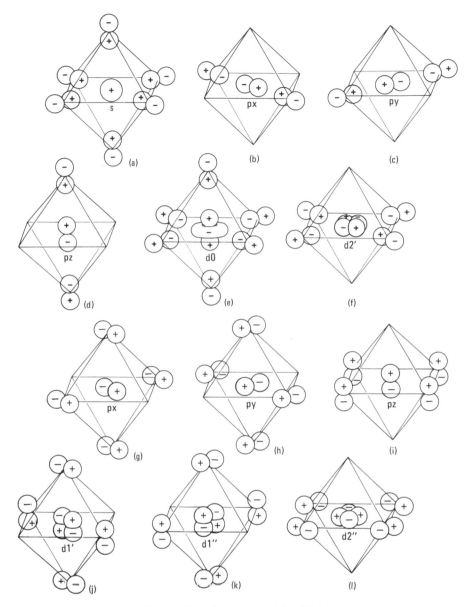

Figure 8-17. SOs for $CoF_6^{3-}$ generated by **GO**s (see Table 8-7).

convenient to maintain the axis system used in the CFT model in which the ligands lie along the axes of the metal center. In Figure 8-17a–f are shown the SOs generated in the radial $2pz$ AO set and in Figure 8-18g–r are depicted the twelve SOs generated in the tangentially oriented $(2px)$ and $(2py)$ VAO set.

The information in Figure 8-17 is summarized in Generator Table 8-7. In

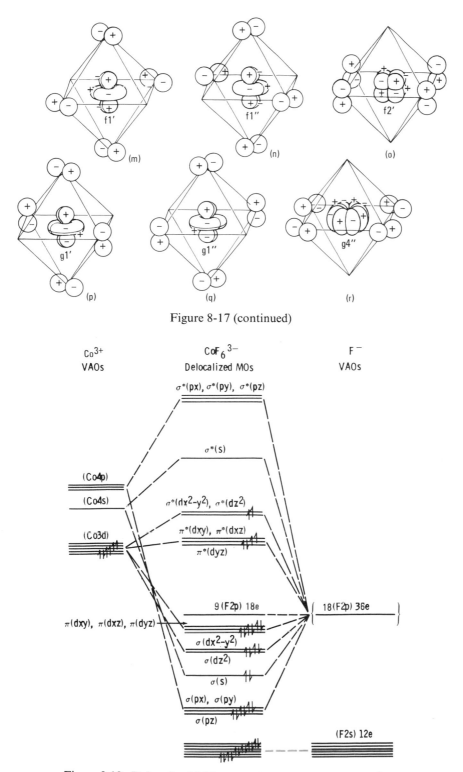

Figure 8-17 (continued)

Figure 8-18. Delocalized MO energy level diagram for $CoF_6{}^{3-}$.

this table we see that the cobalt $(d)$ VAOs are divided into two equivalence sets because they break up into two partner sets in the octahedral potential. Since $F^-$ ion produces a weak field complex and $Co^{3+}$ is a $d^6$ ion, we place each of the first five electrons in separate cobalt $(d)$ VAOs in the crystal field orbital diagram in Figure 8-16 and the sixth electron would pair with an electron in the $(dxy)$, $(dxz)$, $(dyz)$ set. In the covalent delocalized view we are developing, we expect these electrons to be housed in MOs having mainly cobalt $(3d)$ character. From the relative energies of the Co $(3d)$ VAOs and the F $(2p)$ VAOs with which they interact (Figure 8-17), it is seen that the MOs with appreciable cobalt $(3d)$ character are $\sigma^*(dx^2 - y^2)$, $\sigma^*(dz^2)$, $\pi^*(dxy)$, $\pi^*(dxz)$, and $\pi^*(dyz)$.

Alternatively, we can reason that the four ABMOs arising from the linear combination of the $s$-, $px$-, $py$-, and $pz$-generated SOs must lie at higher energies than the ABMOs stemming from the linear combination of any of the $d$-generated SOs, since the cobalt $(4s)$ and $(4p)$ VAOs are at higher energy than the $(3d)$ VAOs. Moreover, splittings of $\pi$ MOs tend to be small compared to $\sigma$ interactions. Leaving the highest four ABMOs empty allows twenty-three MOs to remain for occupation by forty-two electrons after preassignment of two lp's on each fluorine. From the experimental observation that the complex possesses four unpaired electrons, it is clear that the five ABMOs from the $d$-generated SOs must be occupied as shown in Figure 8-18.

Twelve of the remaining thirty-six electrons are placed in the NBMOs which are the SOs generated by $f1'$, $f1''$, $f2'$, $g1'$, $g1''$, and $g4''$. Twelve more electrons are placed in the BMOs obtained by the linear combinations of the pairs SOs generated by the $s$, and $d$ **GO**s. The twelve electrons left must be housed in the remaining six MOs obtained by linearly combining the SOs generated by $px$, $py$, and $pz$. (Recall that three ABMOs from these sets are empty.) We consider first the $px$-generated set. Any two orbitals formed as linear combinations of the occupied canonical MOs can serve as filled $px$-generated MOs. The filled MOs are the lowest-energy MOs, the empty MO being antibonding. We can always choose the pair of occupied orbitals so that one of the SOs does not contribute to one of the MOs. If we choose the $(Co4px)$ SO to be noncontributing, linear combination of the fluorine SOs in Figure 8-17b and g yields a manifestly nonbonding MO since the fluorines are too greatly separated for appreciable interaction among themselves. The same argument also applies to the $py$- and $pz$-generated SOs in Table 8-7. Thus we have identified three NBMO–BMO pairs which house the remaining twelve electrons. The three doubly occupied NBMOs are among the nine-fold isoenergetic levels in Figure 8-18.

The energy difference $\Delta$ as calculated from ligand field theory is, of course, different from that obtained from crystal field theory. The more ionic the interaction is, however, the more the two values will tend to agree. The doubly and triply degenerate $(d)$ orbitals in the CFT model for octahedral systems arise from ionic interactions, whereas in the LFT approach the corresponding triply degenerate MOs are largely nonbonding (partially antibonding) metal VAOs, and the corresponding doubly degenerate MOs possess substantial

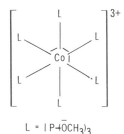

$$L = | P\overline{+}OCH_3)_3$$

Figure 8-19. Electron dash structure for Co[P(OCH$_3$)$_3$]$_6^{3+}$.

antibonding character. Covalency is also evident in the BMOs, in that ligand electrons do attain some metal character owing to orbital overlap in the MOs. Finally, we note that the bond order is difficult to determine because of the partial occupation of $\sigma$ and $\pi$ ABMOs.

Before proceeding to the localized bond picture, it should be noted that ligands such as PH$_3$ or AsF$_3$ have no lone pairs on the ligating Group V atom for $\pi$ interactions with the metal ($d$) VAOs. However, there are empty ($d$) VAOs on phosphorus and arsenic in such ligands, respectively, and there is evidence that such interactions may occur between these orbitals and the metal ($d$) VAOs. The example with which we choose to illustrate this concept is the octahedral, colorless trivalent cobalt complex Co[P(OCH$_3$)$_3$]$_6^{3+}$ whose electron dash structure appears in Figure 8-19. The ($d$) VAOs on the phosphorus which have the correct symmetry for $\pi$ interactions with the metal are the (P3$dxz$), (P3$dyz$), VAOs. By referring to the axis system in Figure 8-12, it is seen that the two lobes of these VAOs which are directed into the octahedron for such interactions have the same symmetry with respect to the octahedron as do the (P3$px$) and (P3$py$) VAOs. Thus the SOs in Figure 8-17g–r are seen to be analogous to those expected for a set of ($dxz$) and ($dyz$) ligand orbitals. The SOs for the (P3$dxy$) AOs are depicted in Figure 8-20. The interactions of these AOs with the ($dxz$), ($dyz$), and ($dxy$) VAOs of cobalt are $\delta$, rather than $\pi$. A $\delta$ interaction is seen to be weaker than $\pi$ and we will view it as negligible in the present instance. From Figures 8-17g–r and 8-20 we deduce that, except for Figure 8-17j–l, all the ligand SOs form NBMOs to a first approximation. Thus the SOs in Figure 8-17g–i form largely NBMOs because the metal (4$p$) VAOs are mainly involved in $\sigma$ bonding. The other SOs which become non-bonding (Figure 8-17m–r) derive this character from the fact that the GOs which generated them possess symmetries which are incompatible with any metal AOs in the valence set. *In general, valence (d) VAOs on ligands lie above the metal (d) VAOs.* Thus the splitting between the resulting triply degenerate BMO and ABMO levels (see Figure 8-17j–l) tends to *increase* $\Delta$.

In formulating a localized approach for CoF$_6^{3-}$, we note in Figure 8-18 that an $s$, three $p$, and five $d$ GOs are associated with the occupied bonding levels in the delocalized MO scheme for CoF$_6^{3-}$. We should also be aware that the Lewis structure of this ion in Figure 8-16 suggests that there will be a set of six localized BMOs, one lp, and four lone singles. We consider first the

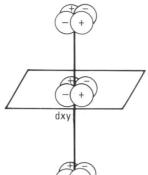

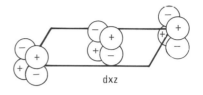

(a)

(b)

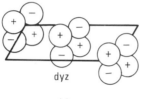

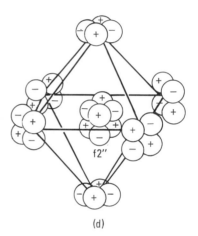

(c)

(d)

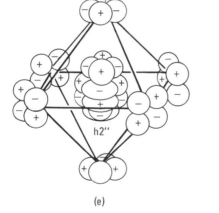

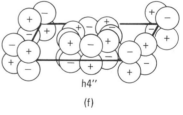

(e)

(f)

Figure 8-20.  SOs for the phosphorus ($3dxy$) AOs in $Co[P(OCH_3)_3]_6^{3+}$.

six delocalized sigma BMOs which are generated by the $s$, the three $p$, the $dz^2$, and $dx^2 - y^2$ **GO**s. These six **GO**s can be hybridized so that the main lobes form an octahedrally oriented $d^2 sp^3$ **GO** set. These hybrid **GO**s generate an identical set of cobalt VAOs and they also call in a ($2pz$) VAO on each fluorine. Six localized MOs can then be formed from the six pairs of cobalt–fluorine VAOs.

The doubly occupied ABMO in Figure 8-18 is composed predominantly of one of the ($d$) VAOs in the triply degenerate set, or some linear combination of these VAOs. We are thus unable to further localize this electron pair. The same is true for the remaining two unpaired electrons in this orbital set and the two in the doubly degenerate set since they reside in MOs which arise from linear combinations of SOs generated by GOs which lie in the same set. Because the $\pi$ BMOs also arise from SOs generated by $dxy$, $dxz$, $dyz$, their electron pairs are not further localized. The electron pairs in the nine NBMOs are already largely localized in fluorine orbitals and so we do not attempt further localization. In Figure 8-21 is an energy level diagram in which the above considerations are reflected.

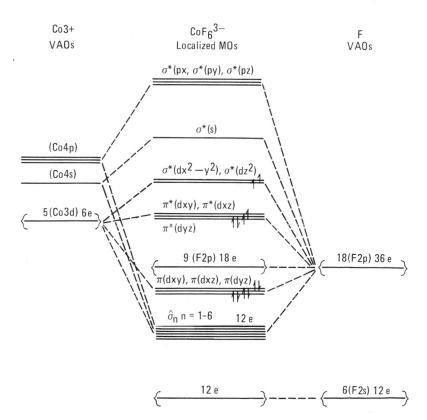

Figure 8-21. Localized MO energy level diagram for $CoF_6^{3-}$.

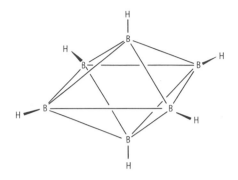

Figure 8-22. Structure of $B_6H_6^{2-}$.

## 8.7. $B_6H_6^{2-}$

Like $B_5H_9$ and other electron-deficient boron hydrides, the structure of the $B_6H_6^{2-}$ ion is not predictable by drawing Lewis structures. We therefore must be satisfied with the experimental finding that $B_6H_6^{2-}$ consists of an octahedral cluster of boron atoms, each possessing a terminal hydrogen as shown in Figure 8-22.

We assign the B—H bond pairs in $B_6H_6^{-2}$ to localized MOs, each of which is composed of an (H1s) and a (B$sp^\alpha$) VAO. The remaining VAOs on each boron are made up of a digonal (B$sp^\beta$) hybrid pointed into the center of the octahedron and a (B2px) and a (B2py) orbital oriented tangentially around the octahedron along the axes shown in Figure 8-12. The SOs in the inward-pointing and tangential sets are depicted in Figure 8-17 and are summarized along with their corresponding GOs in Generator Table 8-8. It may be noted that the equivalence sets in this table are analogous to the F$pr$ and F$pt$ equivalence sets in Table 8-7. Although the GOs are also the same, the entries under them in Table 8-8 reflect the fact that the boron atom VAOs interact with one another. Examination of the SOs in Figure 8-17b–d and g–i shows us that the $p$-generated BMOs are linear combinations dominated by the SOs of the B$pt$ set owing to their proximity on adjacent atoms. In the energy level diagram in Figure 8-23, the electron occupation leads us to conclude that the average bond order is 7/12 per B—B link.

From Figure 8-23 it is seen that seven GOs are associated with the filled BMOs. If we treat the inwardly directed and the tangential MOs separately, we find that, although no localization of the single bond pair in $\sigma(s)$ stemming from the inwardly directed MO is possible, it is possible to localize the six tangentially oriented bonding MO pairs on the basis of the directionality of the GOs associated with the two triply degenerate sets of BMOs. The relevant GO hybrid set is $p^3d^3$, whose members point to the vertices of a trigonal antiprism. Such an antiprism can be visualized in the octahedral array of boron atoms by picking any pair of oppositely oriented triangular faces in

Table 8-8. Generator Table for $B_6$ Cluster in $B_6H_6^{2-}$ (14 e⁻)

| VAO Equivalence Sets | GOs | | | | | | | | | | | | | | |
|---|---|---|---|---|---|---|---|---|---|---|---|---|---|---|---|
| | $s$ | $px$ | $py$ | $pz$ | $dxy$ | $dx^2-y^2$ | $dxz$ | $dyz$ | $dz^2$ | $f1'$ | $f1''$ | $f2'$ | $g1'$ | $g1''$ | $g4''$ |
| $Bsp$-in = $(B_A sp^b)$, $(B_B sp^b)$, $(B_C sp^b)$, $(B_D sp^b)$, $(B_E sp^b)$, $(B_F sp^b)$ | b | n | n | n | | a | | | a | | | | | | |
| $Bpt$ = $(B_A 2px)$, $(B_A 2py)$, $(B_B 2px)$, $(B_B 2py)$, $(B_C 2px)$, $(B_C 2py)$, $(B_D 2px)$, $(B_D 2py)$, $(B_E 2px)$, $(B_E 2py)$, $(B_F 2px)$, $(B_F 2py)$ | | b | b | b | b | | b | b | | a | a | a | a | a | a |

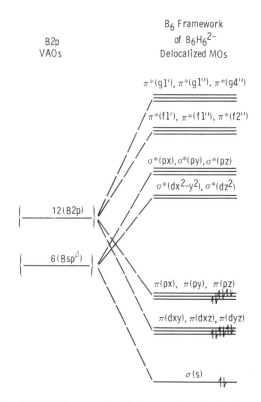

Figure 8-23. Delocalized MO energy level diagram for the $B_6$ framework of $B_6H_6^{2-}$.

the octahedron and imagining the main lobes of the $p^3d^3$ hybrid set to point almost toward the midpoints of the edges of the triangles as indicated in Figure 8-24. Six bp's are localized in the regions toward which these six lobes point. It must be realized that there are antibonding contributions from the other boron $(2px)$ and $(2py)$ VAOs, which confer a bond order of 7/12 on each localized bond. The natures of these antibonding interactions are complicated by the rather complex geometry of the octahedron and we do not further discuss them. The energy level diagram which corresponds to this localized view is shown in Figure 8-25. By localizing the six bp's near the edges of the opposite triangular faces as shown in Figure 8-24, we see that these opposite faces do not appear to be bound to one another. Because the manner in which we orient the trigonal antiprismatic array of bond pairs is arbitrary, however, three additional resonance forms can be drawn which involve the remaining three pairs of opposite triangular faces. Recall that a similar set of $\sigma$-bonded resonance structures was encountered for the electron-deficient ion $H_3^+$. In contrast to the case at hand, wherein some localization can be achieved, no localization was realized for the two electrons available in $H_3^+$.

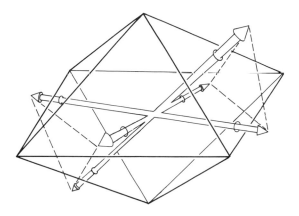

Figure 8-24. Sketch of $p^3d^3$ hybrid **GO** set inscribed in an octahedron.

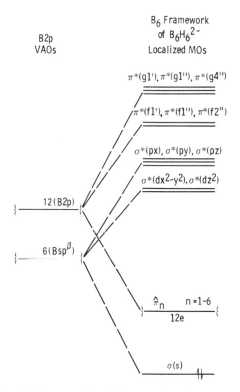

Figure 8-25. Localized MO energy level diagram for the $B_6$ framework of $B_6H_6{}^{2-}$.

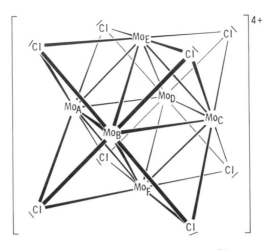

Figure 8-26. Structure of $Mo_6Cl_8^{4+}$.

## 8.8. $Mo_6Cl_8^{4+}$

Reaction of $MoCl_4$ with molybdenum metal at 600°C yields $Mo_6Cl_{12}$ from which four chloride ions are easily lost to give $Mo_6Cl_8^{4+}$. Although a Lewis structure which is not electron deficient can be drawn (Figure 8-26), the prediction of the experimentally determined structure by any set of simple rules is not feasible. Many octahedral metal cluster compounds are known, but other symmetrical geometries, including triangular, tetrahedral, square pyramidal, trigonal bipyramidal, trigonal prismatic, and icosahedral have also been observed, although none of these are currently as numerous as the octahedral examples. In the structure in Figure 8-26, each line represents an electron pair so that the Lewis structure is also represented. Each Cl—Mo bond can be viewed as a localized 2-center 2-electron bond composed of a ($p$) VAO on the chlorine and one member of a square planar set of molybdenum ($spxpydxy$) hybrid VAOs. (The chlorine lone pair is assigned to a (Cl3s) VAO.) It is seen then, that we have twenty-four valence electrons with which to bind the $Mo_6$ cluster. The greater electronegativity of the chlorines imparts considerable ionicity to the Mo–Cl bonds.

Using the axis system in Figure 8-12 which we employed for the bonding treatment of the $B_6H_6^{-2}$ ion, also an octahedral cluster, we construct Generator Table 8-9. In this table the M$opr$ equivalence set is composed of M$opzdz^{2\beta}$ hybrids that point inward while the other members of the digonal pairs (i.e., M$opzdz^{2\alpha}$) are directed outward. Comparing Table 8-9 with Table 8-8, we see that the only difference in **GO** requirements is that imposed by the inclusion in Table 8-9 of a VAO set of the $(dx^2 - y^2)$ type on each cluster atom. The SOs for these VAOs are depicted in Figure 8-27. Notice in this figure how the nodal properties of the **GO** carry over into SOs (MOs) formed by the molybdenum $(dx^2 - y^2)$ VAOs. This is particularly obvious in Figure 8-27b

Table 8-9. Generator Table for $Mo_6$ Cluster in $Mo_6Cl_8^{4+}$ (24 e⁻)

| VAO Equivalence Sets | | | | | | | | GO$s$ | | | | | | | | |
|---|---|---|---|---|---|---|---|---|---|---|---|---|---|---|---|---|
| | $s$ | $px$ | $py$ | $pz$ | $dxy$ | $dx^2-y^2$ | $dxz$ | $dyz$ | $dz^2$ | $f1'$ | $f1''$ | $f2'$ | $g1'$ | $g1''$ | $i2'$ | $g4''$ |
| $Mopr =$ $(Mo_A pzdz^{2\beta})$, $(Mo_B pzdz^{2\beta})$, $(Mo_C pzdz^{2\beta})$, $(Mo_D pzdz^{2\beta})$, $(Mo_E pzdz^{2\beta})$, $(Mo_F pzdz^{2\beta})$ | b | n | n | n | | a | | | a | | | | | | | |
| $Moda =$ $(Mo_A 4dx^2-y^2)$, $(Mo_B 4dx^2-y^2)$, $(Mo_C 4dx^2-y^2)$, $(Mo_D 4dx^2-y^2)$, $(Mo_E 4dx^2-y^2)$, $(Mo_F 4dx^2-y^2)$ | | | | | | b | | | | | | | | | a | |
| $Modb =$ $(Mo_A 4dxz)$, $(Mo_A 4dyz)$, $(Mo_B 4dxz)$, $(Mo_B 4dyz)$, $(Mo_C 4dxz)$, $(Mo_C 4dyz)$, $(Mo_D 4dxz)$, $(Mo_D 4dyz)$, $(Mo_E 4dxz)$, $(Mo_E 4dyz)$, $(Mo_F 4dxz)$, $(Mo_F 4dyz)$ | | | | | | | | | b | n | n | n | | | | |
| | | b | | | b | | b | b | a | a | a | a | a | a | | a |

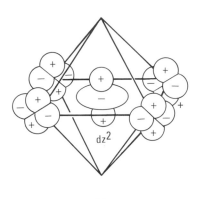

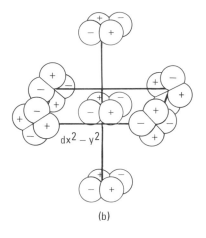

(a)

(b)

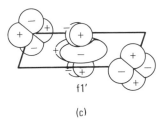

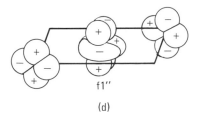

(c)

(d)

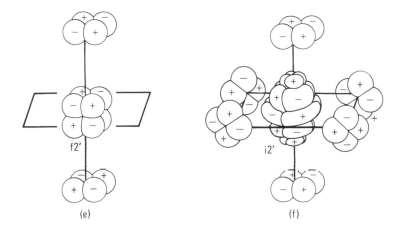

(e)

(f)

Figure 8-27.  SOs for the molybdenum ($4dx^2$-$y^2$) VAOs in $Mo_6Cl_8^{4+}$.

and e. In Figure 8-27a the two cones of the $dz^2$ **GO** call in VAO lobes above and below the plane of the same sign and lobes of the opposite sign all around the edge of this plane. Thus the upper positive lobes are separated from the negative lobes by a cone (whose surface undulates in a regular manner around the $z$ axis). The lower positive lobes have a similar conal node associated with them. The same is true in Figure 8-27c and $d$ except that each **GO** and **SO** also contains a planar node. In Figure 8-27f there are two additional cones as well as another plane. The question may be asked why the additional two cones are necessary. The answer is that if they were absent, the SO generated in Figure 8-27b would be duplicated. The twenty-four electrons available to bind the molybdenum cluster can be accommodated by twelve bonding orbitals, one from each of the first twelve columns of SOs in Table 8-9. As expected from the Lewis structure, the bond order for $Mo_6Cl_8^{4+}$ indicated by the energy level diagram in Figure 8-28 is 1.0.

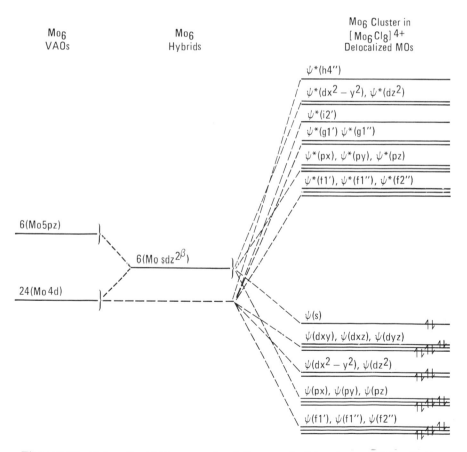

Figure 8-28. Delocalized MO energy level diagram for $Mo_6$ cluster of $Mo_6Cl_8^{4+}$.

Table 8-10. A Useful Summary of Procedures for Generating Pictorial
Representations of Delocalized and Localized Bonding Views of a Molecule and
for Visualizing its Vibrational Modes

| Delocalized View | Localized View | Normal Modes |
|---|---|---|
| 1. Decide on best electron dash structure using lowest possible formal charges and drawing resonance forms. In such forms, some lp's may be involved in a $\pi$ system. | 1. Decide on which set(s) of doubly occupied delocalized MOs to localize. | 1. Identify the motional atomic vector (AV) equivalence sets and list them in the generator table. |
| 2. Establish the molecular geometry from VSEPR theory if possible. *For transition metal complexes and boron hydrides, other considerations apply.* | 2. Hybridize the **GO**s which give rise to the sets(s) in 1. | 2. Using **GO**s, sketch the symmetry motions (SMs). |
| 3. After preassigning lp's to (*s*) and suitably oriented (*p*) VAOs, identify all the VAO equivalence sets (hybrids may also be used) and list them in the generator table. | 3. Using the hybridized **GO**s, sketch the SOs.  4. Sketch the MOs, being aware of the consequences of a central atom on which the full set of SOs (hybridized VAOs) is not generated by the hybridized **GO** set. If necessary, rehybridize peripheral atom VAOs and rotate hybrid **GO**s to locate bp's between atoms. | 3. Fill in the rest of the generator table. (i.e., **GO**s and m entries).  4. Take appropriate linear combinations of SMs generated by the same **GO** and sketch all the molecular motions.  5. Identify the normal vibrational modes. |
| 4. Using **GO**s located at a molecular center or centroid, sketch all the SOs, recalling that #SOs = #VAOs. | 5. Draw the resulting MO energy level diagram and occupy the appropriate levels with electrons. | |
| 5. Fill in rest of generator table (i.e., the necessary **GO**s and the *n*, *a*, or *b* entries). Establish partner **GO** and SO sets. | 6. Verify that the bond order per link is the same as in the delocalized view. | |
| 6. Take linear combinations of SOs generated by the same **GO** and sketch all the MOs. | | |
| 7. Draw an MO energy level diagram and occupy the appropriate levels with electrons. | | |
| 8. Deduce the bond order per link. | | |

The SOs associated with the occupied delocalized MOs can be mixed to form an $sp^3 d^5 f^3$ hybrid **GO** set whose members direct their main lobes to the edges of an octahedron. These **GO** hybrids call in contribution from the (Mo5$pz$) VAO, the (Mo5$s$) VAO, and all the (Mo4$d$) VAOs [except (Mo4$dxy$) which was used to bind the chlorines]. Although the nature of these contributions is complicated, the result is that a bond pair is localized to a large extent at the midpoint of each of the twelve edges of the octahedral metal cluster. This localized picture is in conformity with the Lewis structure.

The molecular motions of $Z_6$ and $YZ_6$ species are obtained in the usual manner. The same is true for a cluster system such as $[MoCl_8]^{4+}$, although it would be a somewhat complicated process. The molecular motions of octahedral $Z_6$ molecules are obtained by taking appropriate linear combinations of the symmetry motions for $Z_6$ represented in Figure 8-17. For $ZY_6$, the SMs of the central atom AVs must be taken into account. Sketching these motions is left as an exercise for the reader.

## Summary

In this chapter square planar $ICl_4^-$, tetragonal pyramidal $BrF_5$, and octahedral $PF_6^-$ were found to be examples of diamagnetic main-group molecules whose central atoms are hypervalent. The transition metal complex $CoF_6^{3-}$, however, was found to be paramagnetic with four unpaired electrons, owing to the participation of ($d$) VAOs in the bonding. In studying this ion, the relevancy of crystal and ligand field concepts was discussed. The probable importance of metal-to-ligand $\pi$ bonding in the related diamagnetic complex $Co[P(OCH_3)_3]_6^{3+}$ was also pointed out. Of the three cluster systems investigated, namely, $B_5H_9$, $B_6H_6^{2-}$, and $Mo_6Cl_8^{4+}$, the first two are electron deficient while the last one is not. In the discussion of $B_5H_9$, the differences between bonds to bridging and terminal hydrogens were elucidated (using $B_2H_6$ as an example) in order to facilitate the bonding treatment of molecules in which such hydrogens occur.

In Table 8-10 is a slightly amended version of our procedures using the **GO** approach. The change reflects the fact that structures of transition metal complexes and metallic and non-metallic cluster compounds are not governed by VSEPR theory.

### PROBLEMS

1. Sketch the SOs for $ICl_4^-$. Do the same with the axis system of the **GO**s rotated by 45° in the $xy$ plane.

2. Develop the delocalized and localized views of the stable compound $XeF_4$.

3. Sketch the molecular motions of $BrF_4^-$ and give the corresponding **GO** table. In your sketches, also show the SMs as combinations of *resultant* AVs on the fluorine atoms.

4. Sketch the delocalized and localized MO energy level diagrams for (a) $ICl_4^-$, (b) $BrF_5$, (c) $PF_6^-$.

5. In the extended array compound $Fe_{1.89}Mo_{4.1}O_7$, square planar oxygens are bound to four Mo atoms. (a) Develop the delocalized view of the $Mo_4O$ unit assuming that each Mo—O $\sigma$ bond involves an electron from each atom and that $\pi$ bonding can only occur from the oxygen to a ($d$) VAO on each Mo. (b) Rationalize why the square planar oxygen–metal bonds in $Fe_{1.89}Mo_{4.1}O_7$ ($\sim 2.15$ Å) are longer than the trigonal planar oxygen–metal links in $NaMo_4O_6$ ($\sim 2.05$ Å).

6. Sketch the axis system used to generate the SOs for $BrF_5$ in Table 8-3. Show the approximate location you would use for the **GO** center if you consider the experimentally determined structure and give a reason for your choice. Sketch the SOs for this structure.

7. Develop the localized view of $BrF_5$ including ($d$) orbital participation.

8. The structure of $InCl_5^{2-}$ is tetragonal pyramidal with the indium atom slightly above the plane of the square array of chlorines. Give a reason for this observation based on VSEPR theory. Is $InCl_5^{2-}$ isoelectronic with $SbCl_5^{2-}$?

9. Account for the fact that the bond lengths decrease on going from $IF_5$ to $XeF_5^-$ (equatorial, 1.869 to 1.845 Å; axial, 1.844 to 1.793 Å) in spite of the fact that Xe is larger than I.

10. Account for the decrease in $F_{ax}XF_{eq}$ bond angle on going from $BrF_5$ (84.8°) to $IF_5$ (81.9°). Both the axial and equatorial bond lengths increase on going from the bromine to the iodine compound.

11. Consider the localized view of the $B_2H_2$ ring of $B_2H_6$. Show that by pointing the main lobes of the $sp$ **GO** hybrids toward diagonally oriented B—H links, one of two resonance forms is generated, in each of which the ring system is broken.

12. Sketch the SOs of $B_5H_9$ using the axis system in Figure 8-9.

13. There is some evidence that the structure below can exist as a transient species. In this cation the apical C—H is presumably bonded via carbon to the $\pi$ orbital system of the cyclobutadiene. Develop the delocalized view of this cation by considering the interaction of the CH fragment with the $\pi$ system of the $C_4$ fragment.

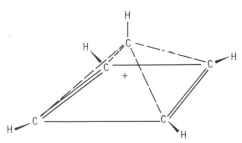

14. Using CFT briefly account for the relative magnitudes of the metal–ligand distance in the following pairs of ions: (a) $FeF_6^{3-}$ (1.90 Å) and $FeF_6^{4-}$ (2.08 Å); (b) $NiF_6^{3-}$ (1.85 Å) and $NiF_6^{4-}$ (2.00 Å)

15. Briefly account for the expectation that there would be no splitting of the $(d)$ orbitals in the $TiF_6^{2-}$ ion if the interactions were purely ionic, but that there would be splitting if any covalency were present.

16. Using CFT show how the $(d)$ orbital energy levels change when the ligands on the $z$ axis are compressed in an octahedral complex. Show how such a distortion could relieve the degeneracy in (a) a weak-field $d^4$ complex; (b) a weak-field $d^9$ complex; (c) a strong-field $d^9$ complex.

17. By means of an energy level diagram, show that the presence of empty $(d)$ VAOs on the phosphorus in diamagnetic $Co\,[P(OCH_3)_3]_6^{3+}$ can increase $\Delta$ over the value expected in the absence of such $\pi$ bonding. What is the bond order?

18. Several years ago, Nobel laureate G. Wilkinson reported the octahedral compound $W(CH_3)_6$. Formulate the delocalized approach assuming the carbon atom uses $sp^3$ orbitals. Develop the localized view using an appropriate set of hybrids. Comment on the magnetism of the Re analogue which was also reported. Which compound would you suppose is more easily oxidized and why?

19. The Cr in $Cr\,(CO)_6$ is believed to back-donate electron density from its $(dxy)$, $(dxz)$, $(dyz)$ VAOs to the $\sigma^*$ MOs of the CO ligands. Generate the SOs for these interactions and sketch an MO energy level diagram for the $CrC_6$ moiety. Indicate from this diagram how the CrC multiple bonding which results stabilizes the system and why a higher CO stretching frequency is expected in the absence of the $\pi$ back-bonding.

20. Sketch the localized energy level diagram for $Mo_6Cl_8^{4+}$.

21. R. E. McCarley, an American chemist, has succeeded in isolating heterometallic clusters of the type $Ta_5MCl_{12}^{3+}$ and $Ta_5MCl_{12}^{4+}$, where M is Mo or W, and also $Ta_4Mo_2Cl_{12}^{4+}$. (a) Which species are paramagnetic? (b) Which species are isoelectronic? (c) Sketch the isomers possible for the $Ta_4Mo_2Cl_{12}^{4+}$ ion.

22. Reaction of bromine with $[W_6Br_8]Br_4$ gives $[W_6Br_8]Br_6$, which apparently retains the same cation structure as the starting material. In contrast, reaction of chlorine with $[W_6Cl_8]Cl_4$ gives $[W_6Cl_{12}]Cl_6$, in which the cation has the structure shown below. Sketch the delocalized MO diagrams for the cations in both tungsten products.

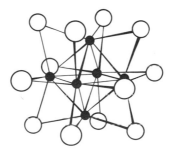

23. Develop the delocalized view of the neutral octahedral nickel cluster of $(\eta^5 - C_5H_5)_6Ni_6$, in which each $\eta^5 - C_5H_5^-$ ion can be considered to donate

three pairs of $\pi$ electrons to three ($sd^3$) metal hybrids obtained by hybridizing ($4s$) with ($3dxy$), ($3dxz$), and ($3dyz$). Such a hybrid set possesses main lobes that point to the cornérs of a tetrahedron. Why is it better to pick an $sd^3$ rather than an $sp^3$ hybrid set?

24. Show that by pointing the **GO** axis sytem to the atom positions in $CoF_6^{3-}$ that the $d$-generated SOs involving the tangential ($p$) VAOs on the fluorines do not readily reveal the partner relationships among the cobalt (d) VAOs.

25. Sketch the molecular motions of $BrF_5$ and of $CoF_6^{3-}$.

# CHAPTER 9

# Tetrahedral and Related Molecules

In this chapter we consider $P_4$ which consists of a tetrahedral cluster of like atoms. Other examples in this class are the isoelectronic $Tl_4^{8-}$, $Si_4^{4-}$, $Ge^{4-}$, $Pb^{4-}$, $As_4$, and $Te^{4+}$. We will also examine $PO_4^{3-}$ and $CH_4$ in which a central atom is present within the tetrahedral array of peripheral atoms. Additional examples in this broad class of molecules are transition metal complexes such as $CrO_4^{3-}$, $CoCl_4^{2-}$, $VCl_4^-$, $FeCl_4^-$, and $TiCl_4$. Since transition metal complexes possess interesting bonding features, we will briefly study $VCl_4^-$. We begin, however, with $PnH_3$ (Pn = pnicogen = N, P, As, Sb, Bi) whose structure is trigonal pyramidal and can be considered as tetrahedral with one vertex atom missing.

## 9.1. $PnH_3$

The first three members of the pnicogen hydrides are thermally stable gases, while $SbH_3$ and $BiH_3$ easily decompose to the constituent elements. From the normal electron dash structure of $PnH_3$ in Figure 9-1 and from VSEPR considerations we infer that these molecules are pyramidal. The experimentally determined bond angles tend to be near 90° except for $NH_3$ in which this angle is 107°. The larger angle in the latter compound can be rationalized on the basis of the fact that nitrogen is the smallest pnicogen and therefore the bp–bp repulsions assume greater importance. In addition, the energy separation between $(ns)$ and $(np)$ VAOs increases with $n$ (owing to greater nuclear shielding effects) so that $sp$ mixing is less likely in the heavier congeners. Mixing of $(s)$ and $(p)$ VAOs is typical in elements of the second row and we have already seen an example of such behavior in $H_2O$, which possesses a wider bond angle than its congeners $H_2S$ or $H_2Se$.

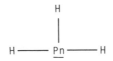

Figure 9-1. Electron dash structure for $PnH_3$ where Pn = N, P, As, Sb, and Bi.

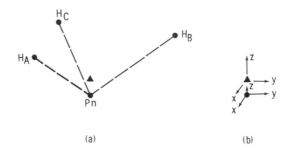

(a)                                          (b)

Figure 9-2. Geometry of $PnH_3$ (a) and axis system for **GO** center and Pn orbitals (b).

Table 9-1. Generator Table for $PnH_3$ (6 e$^-$)

| VAO Equivalence Sets | | GOs | |
|---|---|---|---|
| | $s$ | $px$ | $py$ |
| Pn $pt$ = (Pn $npx$), (Pn $npy$) | | n | n |
| Pn $pr$ = (Pn $npz$) | n | | |
| H$s$ = (H$_A$1$s$), (H$_B$1$s$), (H$_C$1$s$) | n | n | n |

After reserving an $ns$ VAO on the pnicogen for the lp, the delocalized MOs are obtained in the usual manner (see Figure 9-2 and Generator Table 9-1). Two of the BMO–ABMO pairs are shown as degenerate in the energy level diagram in Figure 9-3. This degeneracy, which arises from the fact that the SOs are generated by the $px$ and $py$ **GO**s, is analogous to that discussed in detail for the case of $H_3^+$. As expected from the electron dash structure, the bond order is 1.0.

Since the occupied BMOs in Figure 9-3 are associated with the **GO** set $s$, $px$, and $py$, the hybrid **GO** set $sp^2$ is required for the localized view. This hybrid **GO** set can be oriented on the **GO** center such that the main lobe of each member points to a line connecting a hydrogen and the pnicogen as shown in Figure 9-4. The three bonding pairs resulting from this localization form 2-center bonds composed of an (H1$s$) VAO and a (Pn$p$) VAO. These ($p$) VAOs are directed toward the (H1$s$) lobes, as shown for one of them in Figure 9-4, by simply rehybridizing them from the orientation they possess in our delocalized view (Figure 9-1). In the localized energy level diagram in Figure 9-5 we see that the average bond order (1.0) is the same as that determined in our delocalized view.

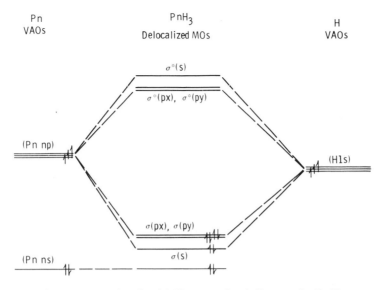

Figure 9-3. Delocalized MO energy level diagram for PnH$_3$.

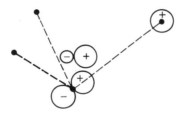

Figure 9-4. The orientation of one $sp^2$ **GO** hybrid and the VAOs it calls in on one hydrogen and Pn in PnH$_3$.

Experimentally, NH$_3$ is found to possess a bond angle of 107° which is nearly the tetrahedral angle. In the localized view of this molecule, more favorable overlap in the 2-center N—H bonds can be achieved by mixing the (N2s) and (N2p) VAOs so that the hybrids pointing toward the (H1s) lobes are separated by 107° instead of 90°. Such rehybridization, which results in a nearly ($sp^3$) VAO set on nitrogen, places almost 25% ($p$) character in the fourth hybrid in order to preserve approximate orthogonality of the nitrogen contributions to the localized MOs. This fourth hybrid, which points down the negative $z$ axis of the nitrogen in the coordinate system of Figure 9-2b, houses the lone pair. A point to be appreciated from this discussion is that the reservation of a pure (N2s) VAO for the lp is only an approximation and that in the present case the lone pair actually possesses a somewhat higher energy than that expected for a pure (N2s) VAO since there is considerable ($2p$) character mixed with it. By not initially assigning the lone pair to an (N2s)

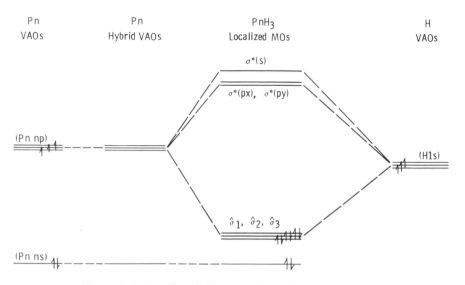

Figure 9-5. Localized MO energy level diagram for $PnH_3$.

VAO, this orbital is introduced into Table 9-1 as an equivalence set. Such a procedure is easily seen to give three SOs generated by an $s$ **GO**, which linearly combine to form an NBMO and a BMO–ABMO pair. Although the (N2$s$) VAO is expected to contribute substantially to the NBMO, a considerable contribution is also expected from (N2$pz$) in view of the ($sp$) mixing which was concluded to occur in the localized view. Thus the lone pair in the delocalized view of $NH_3$ is expected to be higher in energy than an (N2$s$) VAO because of a gain in ($p$) character. In that case we could have reserved an approximately ($sp^3$) hybrid orbital for the lone pair on the nitrogen in $NH_3$ at the outset. Of course, the increase in lp energy due to hybridization with the ($p$) VAO is offset by a decrease in lp–bp repulsions.

The molecular motions are obtained from the **GO**-generated SMs of the AVs and this is left as an exercise.

## 9.2. $P_4$

Elemental phosphorus exists in more than ten allotropic forms. The structures of most of these are still undetermined. For $P_4$, three electron dash structures that obey the octet rule can be drawn as shown in Figure 9-6. In contrast to second-row elements, those in the third row and below do not readily form $\pi$ bonds among themselves. This could arise because these atoms are larger and their ($p$) orbitals, which are in higher valence shells, are more diffuse and consequently are less able to overlap effectively. Therefore, the tetrahedral structure becomes the one of choice on this basis (as well as on formal charge

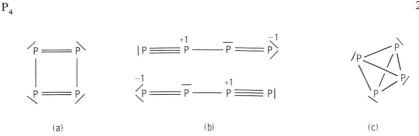

Figure 9-6. Electron dash structures for cyclic (a), chain (b), and cage (c) forms of $P_4$.

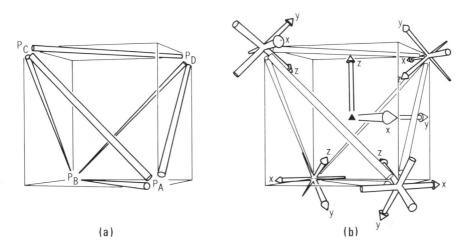

(a)                                                                         (b)

Figure 9-7. View of $P_4$ molecule (a) and the axis system adopted for it (b).

arguments in the case of Figure 9-6b) and indeed it matches the experimentally determined configuration.

The spatial orientation of the $P_4$ molecule is shown in Figure 9-7a and the axis system shown in Figure 9-7b is convenient to use. For a better appreciation of the perspective with which we are viewing this molecule, it has been inscribed inside a cube in this figure. As is usual in our delocalized views, one axis of each peripheral atom points to the center of the molecular species, regardless of whether there is a central atom or not. In the present case, the molecular center is also the **GO** center and this is true in all cases where the **GO** center is unique. Although following our general practice of assigning lone pairs to ($s$) VAOs is, of course, permissible, it is not mandatory. A more precise picture results from not making such a designation, but the resulting analysis is more difficult. Here we choose not to assign lone pairs to (P3$s$) VAOs for pedagogical reasons which will become clear later in the discussion. Among the VAO equivalence sets shown in Generator Table 9-2, it should be noted that *all eight* tangential ($p$) VAOs are in *one* equivalence set. That this must be so is realized, for example, by looking down an axis which goes

Table 9-2. Generator Table for $P_4$ (20 e⁻)

| VAO Equivalence Sets | GOs | | | | | | | | |
|---|---|---|---|---|---|---|---|---|---|
| | $s$ | $px$ | $py$ | $pz$ | $dxy$ | $dz^2$ | $f3'$ | $f3''$ | $f2''$ |
| $Ps = (P_A3s), (P_B3s), (P_C3s), (P_D3s)$ | b | a | a | a | | | | | |
| $Ppr = (P_A3pz), (P_B3pz), (P_C3pz),$ | | | | | | | | | |
| $\quad (P_D3pz)$ | b | a | a | a | | | | | |
| $Ppt = (P_A3px), (P_A3py), (P_B3px),$ | | | | | | | | | |
| $\quad (P_B3py), (P_C3px), (P_C3py),$ | | | | | | | | | |
| $\quad (P_D3px), (P_C3py)$ | | b | b | b | b | b | a | a | a |

through a corner of the tetrahedron and the center of the opposite face. It is apparent that the two tangential ($p$) VAOs on that corner bear the same relationship to each other and to this three-fold axis as the (B2$px$) and (B2$py$) VAOs have with respect to the three-fold central axis perpendicular to the molecular plane of BF$_3$.

The SOs generated among the ($pz$) VAOs are depicted in Figure 9-8 a–d. It is evident from Figure 9-8a that the SO generated by $s$ is bonding. None of the three $p$-generated SOs contains the full symmetry of the molecule and therefore they are three degenerate partner MOs. If we were to sum the squares of the wave functions of three mutually orthogonal ($p$) orbitals, a spherical density would be produced which, of course, contains the full symmetry of any polyhedron, including the tetrahedron. The two members of the $p$-generated SO set shown in Figure 9-8b and c are easily seen to be antibonding. That the $pz$-generated **GO** in Figure 9-8d is antibonding (and it must be since it is degenerate with the $px$- and $py$-generated SOs) can be rationalized by noting that the antibonding interactions between any two adjacent ($pz$) VAOs outnumber the bonding interactions by 2 to 1. The SOs in the P$s$ equivalence set are similarly generated and this information is summarized in Generator Table 9-2.

The pictorial procedure for generating the eight SOs of the P$pt$ equivalence set is shown in Figure 9-8e–s. The $s$ **GO** clearly generates no SO in this set. The three $p$ **GO**s generate three degenerate partner SOs, because none of these SOs contains the symmetry of the molecule. Of the $d$ **GO** set, only $dxy$ and $dz^2$ are used since $dx^2 - y^2$, $dxz$, and $dyz$ duplicate the $pz$-, $px$-, and $py$-generated SOs, respectively. This follows from the fact that the $dx^2 - y^2$- and the $pz$-generated SOs are manifestly the same. Consequently, there *must* exist two other $d$-generated SOs which, together with the $dx^2 - y^2$-generated SO, form the same degenerate partner set as the $px$-, $py$-, and $pz$-generated SOs. By inspection of Figure 9-8 it is readily seen that the SOs needed as partners for the $dx^2 - y^2$-generated SO must be those generated by $dxz$ and $dyz$. The remaining two SOs generated by $d$ **GO**s stem from $dxy$ and $dz^2$. Since neither contains the symmetry of the molecule they too form a degenerate partner set.

There remain three P$pt$ SOs to be found and they are generated by $f$ **GO**s. To begin with, we note that the lobes of the $f2'$ **GO** face directly toward the

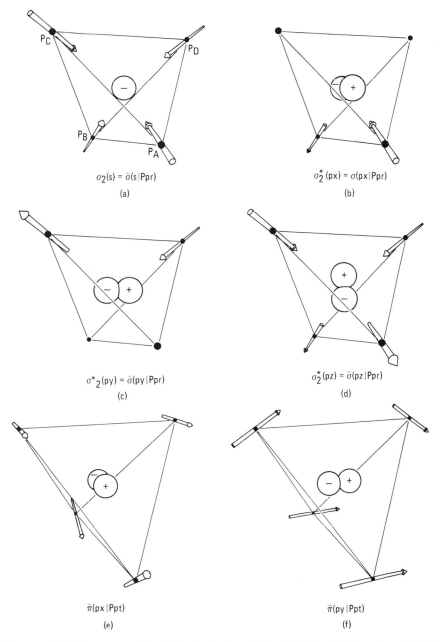

Figure 9-8. SOs generated by **GO**s for P$_4$ (see Table 9-2). The V$\Lambda$O vectors labeled with a question mark [(n), (o)] are discussed in the text. The arrowhead of (P$_B$pt) in (o) is obscured by the lower lobes of the **GO**. The sigma MOs are subscripted with a 2 to differentiate them from the sigma MOs generated in the P$s$ equivalence set (see text). The perspective of the tetrahedron in (a)–(d) is slightly different from (e)–(s) for clarity.

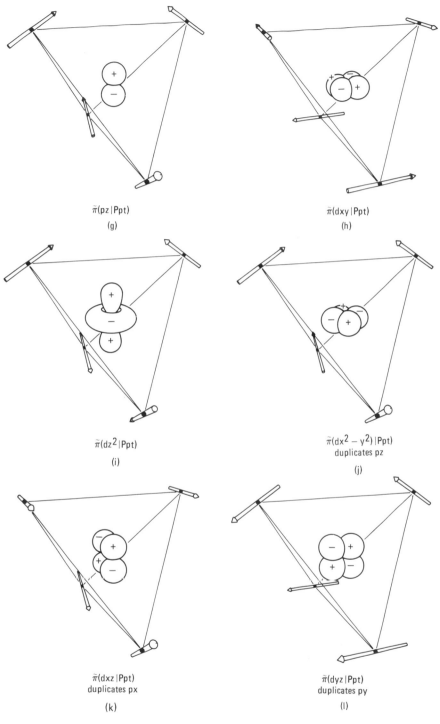

$\tilde{\pi}(\text{pz} \,|\, \text{Ppt})$

(g)

$\tilde{\pi}(\text{dxy} \,|\, \text{Ppt})$

(h)

$\tilde{\pi}(\text{dz}^2 \,|\, \text{Ppt})$

(i)

$\tilde{\pi}(\text{dx}^2 - y^2) \,|\, \text{Ppt})$
duplicates pz

(j)

$\tilde{\pi}(\text{dxz} \,|\, \text{Ppt})$
duplicates px

(k)

$\tilde{\pi}(\text{dyz} \,|\, \text{Ppt})$
duplicates py

(l)

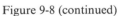

Figure 9-8 (continued)

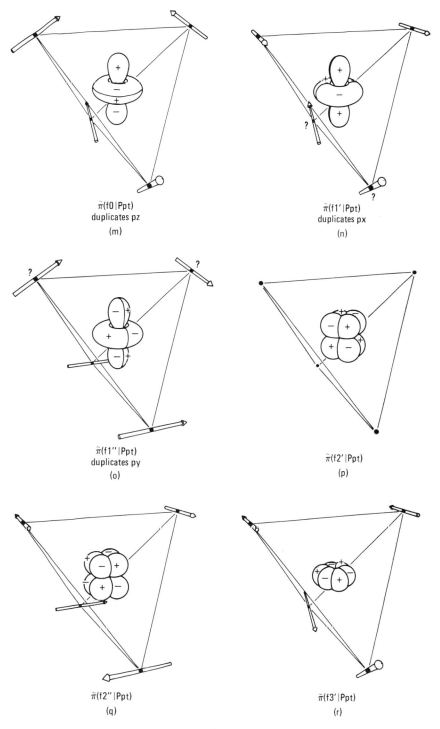

$\tilde{\pi}(\text{f0}\,|\,\text{Ppt})$
duplicates pz
(m)

$\tilde{\pi}(\text{f1}'\,|\,\text{Ppt})$
duplicates px
(n)

$\tilde{\pi}(\text{f1}''\,|\,\text{Ppt})$
duplicates py
(o)

$\tilde{\pi}(\text{f2}'\,|\,\text{Ppt})$
(p)

$\tilde{\pi}(\text{f2}''\,|\,\text{Ppt})$
(q)

$\tilde{\pi}(\text{f3}'\,|\,\text{Ppt})$
(r)

Figure 9-8 (continued)

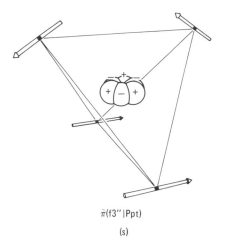

$\bar{\pi}(f3''|Ppt)$

(s)

Figure 9-8 (continued)

corners of the tetrahedron. This **GO** cannot, therefore, generate an SO among the tangential orbitals on phosphorus. Next we observe that the $f0$-generated SO is mainfestly the same as that generated by $pz$ (or $dx^2 - y^2$). From this observation we infer that there must exist two other $f$-generated SOs which complete the partner set to which the $f0$-generated SO belongs. In examining the SOs generated by $f1'$ and $f1''$, the signs with which two of the phosphorus tangential ($p$) VAOs in each SO are called in are not immediately obvious. Thus if the angle subtended by the conal nodes is larger than 109°28', the lower pair of lobes will dominate the manner in which the nearby tangential ($p$) VAOs will be called in in Figure 9-8n (and similarly for the upper pair of lobes in Figure 9-8o) and the arrowheads would be placed as shown. On the other hand, if the conal angle is less than 109°28', each of the VAOs would be more strongly influenced by a "doughnut" segment and an adjacent lower (Figure 9-8n) or upper (Figure 9-8o) lobe. In this case, the question-marked arrowheads would have to be reversed. We can decide this question in two ways. The most straightforward approach is to calculate the angle $\theta$ (see Figure 2-1) from that part of the wave function which governs this angle. In Appendix IV it is easily shown that $\theta = 63°26'$. Thus the total angle of the cone is 126°52', which is larger than the tetrahedral angle of 109°28'. We conclude, therefore, that the arrows are correct as shown in Figure 9-8n and o. We also note that the $f1'$ and $f1''$-generated SOs in these figures duplicate the $px$ and $py$-generated SOs in Figure 9-8e and f. There is a second way in which we could have reached the same conclusion without calculating $\theta$. It is quite clear that the $f2''$, $f3'$, and $f3''$ SOs are new ones as shown in Figure 9-8q–s. From this observation we conclude that the $f1'$-and $f1''$-generated SOs must duplicate SOs generated by simpler **GO**s. The only **GO**s for which this could possibly be true are the $px$ and $py$ orbitals, respectively. These SO duplications require that the phosphorus VAOs labeled with question marks in Figure 9-8n and o must point as shown.

From Generator Table 9-2 we see that the SOs generated by the **GO**s $dxy$, $dz^2$, $f3'$, $f3''$, and $f2''$ are also MOs of the system. The $f2''$-generated SO is manifestly antibonding as seen from Figure 9-8q and the same is true for its partner SOs generated by $f3'$ and $f3''$. Similarly, Figure 9-8i reveals that the $dz^2$-generated SO and hence its partner, the $dxy$-generated SO, are bonding. None of the remaining eleven SOs generated by $s$, $px$, $py$, and $pz$ is an MO by itself. However, all have bonding or antibonding character because they are constructed from VAOs on neighboring atoms. Hence, as was discussed in detail for $N_3^+$, we anticipate that each MO will have a predominant contribution from one SO only, and we can therefore assume that the character of the MO is determined by that of its corresponding SO. From inspection of Figure 9-8 both $s$-generated SOs (MOs) are seen to be bonding. The degenerate $px$-, $py$-, $pz$-generated SOs (MOs) are bonding for the P$pt$ equivalence set but antibonding for the P$s$ and P$pr$ sets. Furthermore, the $(3s)$ AO orbital energy in the free P atom lies some 8 eV below the $(3p)$ AO energy and hence we expect that the *antibonding* as well as the bonding SOs (MOs) of the P$s$ set lie below all other MOs because the remaining MOs are all dominated by $(p)$-type VAOs. Again, the situation is very similar to that in $N_3^+$.

Now, P$_4$ has twenty valence electrons and sixteen MOs. Thus in the ground state the six MOs of highest energy are empty. These are clearly the six ABMOs of predominantly $(p)$ character, namely, the three $f$-generated ABMOs and the three $p$-generated ABMOs (SOs) in the P$pr$ set. The energy level diagram in Figure 9-9 was obtained by means of a calculation and it is seen that our qualitative considerations are reasonable.

The occupied MOs in Figure 9-9 arising from the P$s$ equivalence set are localized by the $sp^3$ hybrid **GO** set whose members have their main lobes directed toward the phosphorus atoms. This results in four localized MOs, each of which consists primarily of a (P3$s$) VAO on a P atom with small antibonding contributions of equal magnitude from each of the three others. These antibonding contributions give rise to nonbonded repulsions among the atoms, which are associated with the requirement of orbital orthogonality and the fact that the $(3s)$ VAOs of neighboring atoms have a non-negligible overlap. If we had reserved the (P3$s$) VAOs for lone pairs at the outset and omitted them from the bonding discussion, we would in fact have been neglecting these nonbonded repulsions. This would not have changed our *qualitative* picture of the molecular orbital scheme but it would have neglected an energy contribution which cannot be omitted for a *quantitative* understanding and which, in fact, might have been important if the bonding effects had been less strong. It should be appreciated that although the (P3$s$) VAOs can be preassigned to contain the lone pairs, this is not because these orbitals do not interact, but rather because the bonding and antibonding MOs of the P$s$ equivalence set are *all* filled.

The remaining occupied MOs in Figure 9-9 are generated by an $s$, three $p$,

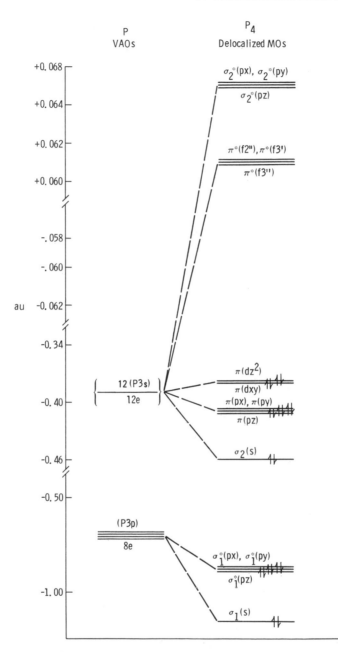

**Figure 9-9.** Delocalized MO energy level diagram for $P_4$. An au is equal to 27.2116 eV.

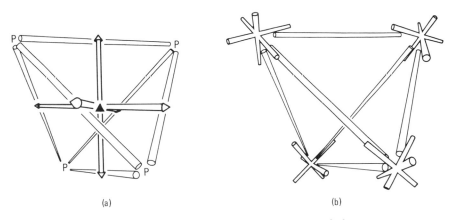

Figure 9-10. Octahedral orientation of the main lobes of the $sp^2 d^2$ **GO** set in P$_4$ (a). In (b) are shown the phosphorus ($p$) VAO sets reoriented to be most effectively called in by the $sp^3 d^2$ **GO** set to form 2-center bonds over the edges of P$_4$.

and two $d$ **GO**s. These **GO**s hybridize to form an $sp^3 d^2$ hybrid **GO** set, whose members point toward the vertices of the octahedron formed by the midpoints of the edges of the P$_4$ tetrahedron as shown in Figure 9-10a. This yields six localized BMOs, each of which essentially connects two P atoms by a 2-center 2-electron bond. The effect of the hybrid **GO** system is to reorient the ($p$) VAOs on each P atom as shown in Figure 9-10b. From this figure we see that the two VAOs which combine to form the bonding lobes of a particular localized BMO do not point directly at each other and thus form bent bonds. The localized energy level diagram in Figure 9-11 satisfies the expectation from the electron dash structures that the average bond order between phosphorus atoms is 1.0.

The localized view of P$_4$ that we have developed is a very satisfying one in spite of the fact that we were forced to use atomic orbitals which interact at an angle. The bent bonds in this molecule are often cited as the source of the observation that P$_4$ is a very highly reactive substance, forming phosphorus compounds in which more normal (less strained) bond angles occur. In fact most of the other forms of elemental phosphorus are much more stable, presumably because they possess bond angles which substantially exceed 60°.

## 9.3. CH$_4$

The tetrahedral shape of CH$_4$ is easily predicted from its Lewis structure (Figure 9-12) and VSEPR considerations. Using the axis system for P$_4$ in Figure 9-7b we can construct Generator Table 9-3. The H$s$ SOs are analogous to the P$s$ SOs in P$_4$. Linear combinations of the pairs of SOs generated by the

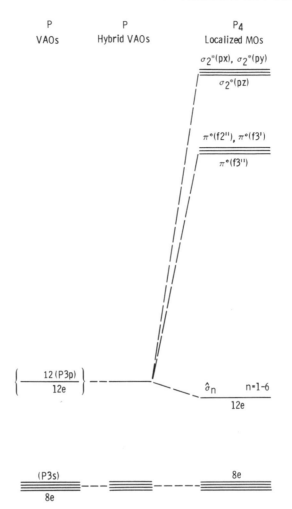

Figure 9-11. Localized MO energy level diagram for $P_4$. Note that the $\sigma$ and $\pi$ occupied MOs in Figure 9-9 are given the designation $\hat{\sigma}$ in the present figure. Although the distinction between $\sigma$ and $\pi$ among orbitals which overlap at an angle is somewhat arbitrary, the localization of an electron pair between the lobes of two such orbitals has the appearance more of $\sigma$ than $\pi$.

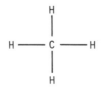

Figure 9-12. Lewis structure for $CH_4$.

Table 9-3. Generator Table for CH$_4$ (8 e$^-$)

| VAO Equivalence Sets | GOs | | | |
|---|---|---|---|---|
| | $s$ | $px$ | $py$ | $pz$ |
| $Cs = (C2s)$ | n | | | |
| $Cp = (C2px), (C2py), (C2pz)$ | | n | n | n |
| $Hs = (H_A1s), (H_B1s), (H_C1s), (H_D1s)$ | n | n | n | n |

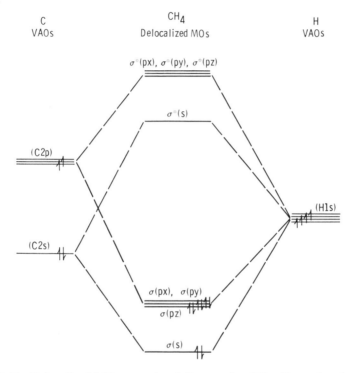

Figure 9-13. Delocalized MO energy level diagram for CH$_4$. (Reproduced with permission from the Journal of Chemical Education.)

same **GOs** give rise to four 5-center delocalized BMO–ABMO pairs whose energy levels are depicted in Figure 9-13. The eight valence electrons occupy the four BMOs, telling us that the average bond order is 1.0.

The occupied BMOs in Figure 9-13 were generated from four **GOs** which, when hybridized, form an $sp^3$ set. This **GO** set generates a corresponding tetrahedral $sp^3$ set of carbon SOs in the VAO set, each member of which is directed toward the corresponding (H1s) VAOs. The resulting 2-center MOs are all equivalent and degenerate. Sketching the energy level diagram is left as an exercise.

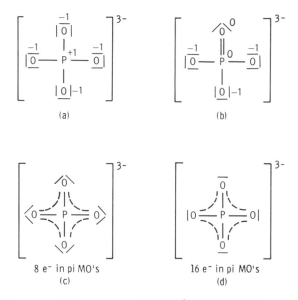

Figure 9-14. Electron dash structures for $PO_4^{3-}$ [(a),(b)]. Average electron dash structures for eight (c) and sixteen (d) electrons in the $\pi$ MOs are also shown.

## 9.4. $PO_4^{3-}$

The electron dash structures for $PO_4^{3-}$ given in Figure 9-14 and VSEPR rules lead us to predict that the ion should be tetrahedral, and this prediction has been confirmed experimentally. On the basis of formal charges, the structure in Figure 9-14b is favored. Moreover this structure has three equivalent partners (which are not shown in Figure 9-14) and together these four resonance structures yield the average resonance form in Figure 9-14c. Provided we have a sufficient number of VAOs on phosphorus, we could incorporate an additional oxygen lp in the $\pi$ system as depicted in Figure 9-14d. By involving all five (P3d) VAOs we will see that this is possible. Although oxygen is more electronegative than phosphorus, the electronegativity difference is not as great as that which occurs in metallic oxides where ionic interactions are of greater importance. Thus the resonance structures shown in Figure 9-15 do not play as important a role in $PO_4^{3-}$.

Since the experimentally determined P—O distance is shorter than that expected for a single bond, some $\pi$ bonding undoubtedly occurs. To permit $\pi$ bonding we see that more than four VAOs on phosphorus will be needed. This observation implies that we could utilize (P3d) VAOs even though their involvement requires an investment of energy.

After reserving the (O2s) VAOs for lp's in the resonance structure in Figure

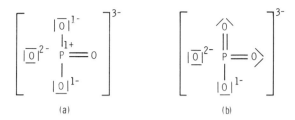

Figure 9-15. Two ionic resonance forms of $PO_4^{3-}$.

9-14d, we construct Generator Table 9-4 from the SOs in Figure 9-8. Because the (P3$d$) AOs are included in the phosphorus VAO set, we must, when taking linear combinations of SOs, consider the fact that the SOs generated by $dx^2 - y^2, dxz, dyz$ duplicate those generated by $px, py, pz$. For simplicity, we shall assume that the $\pi$ interactions between the phosphorus and oxygens involve the (P$d_2$) and (P$d_3$) VAO equivalence sets rather than the P$p$ set. The subscripts in P$d_3$ and in P$d_2$ in Table 9-4 denote the number of (3$d$) VAOs in the equivalence set. Thus we have omitted the $px$-, $py$-, $pz$-generated O$p\pi$ SOs in Table 9-4. We also assume that the P—O $\sigma$ interactions include only the VAOs in the P$s$ and the P$p$ equivalence sets. This simple approach ignores contributions to both the $\pi$ and $\sigma$ systems from the P$p$, P$d_2$, and P$d_3$ equivalence sets which indeed do occur. The twenty-four electrons we have available for the MO system occupy the nine BMOs of the BMO–ABMO pairs and the three $f$-generated NBMOs, as shown in the MO energy level diagram in Figure 9-16. Thus there are sixteen electrons in the $\pi$ system (ten in $\pi$ BMOs and six in $\pi$ NBMOs) and eight electrons in the $\sigma$ system, in accord with the average Lewis structure in Figure 9-14d in which the average bond order is 2.25.

It is instructive at this point to note that the splitting of the triply degenerate $\pi$ BMO–ABMO pairs in Figure 9-16 is greater than that of the doubly degenerate $\pi$ BMO–ABMO pairs. This is most easily visualized by considering an octahedron and its axis system, shown in Figure 8-12, and placing the four oxygen atoms of $PO_4^{3-}$ on triangular faces of the octahedron, as shown in Figure 9-17. In the latter figure the oxygens are at the corners of a tetrahedron which can be inscribed in the octahedron. Notice that as a result of the axis system we have consistently chosen for the octahedron (Figures 8-12 and 9-17) the axis system in the inscribed tetrahedron in Figure 9-17 is rotated by 45° from that used in our treatments of tetrahedral $P_4$ and of $PO_4^{3-}$ (Figure 9-7b). Because the geometrical relationships of the ($d$) orbitals are easier to see in the octahedral axis system we have been using, let us retain the axis system in Figure 9-17. From this figure it is easy to see that the orbitals on any given oxygen are closer to the lobes of the ($dxy$), ($dxz$), ($dyz$) partner set than those of the ($dx^2 - y^2$), ($dz^2$) partner set, since the lobes of the former are directed to the edges of the octahedron while those of the latter are pointed to its vertices.

Table 9-4.  Generator Table for $PO_4^{3-}$ (24 e$^-$)

| VAO Equivalence Sets | GOs | | | | | | | | | | | |
|---|---|---|---|---|---|---|---|---|---|---|---|---|
| | $s$ | $px$ | $py$ | $pz$ | $dxy$ | $dz^2$ | $dx^2-y^2$ | $dxz$ | $dyz$ | $f3'$ | $f3''$ | $f2''$ |
| P$s$ = (P3$s$) | n | | | | | | | | | | | |
| P$p$ = (P3$px$), (P3$py$), (P3$pz$) | | n | n | n | | | | | | | | |
| P$d_2$ = (P3$dxy$), (P3$dz^2$) | | | | | n | n | | | | | | |
| P$d_3$ = (P3$dx^2-y^2$), (P3$dxz$), (P3$dyz$) | | | | | | | n | n | n | • | | |
| O$p\sigma$ = (O$_A$2$pz$), (O$_B$2$pz$), (O$_C$2$pz$), (O$_D$2$pz$) | n | n | n | n | | | | | | | | |
| O$p\pi$ = (O$_A$2$px$), (O$_A$2$py$), (O$_B$2$px$), (O$_B$2$py$), (O$_C$2$px$), (O$_C$2$py$), (O$_D$2$px$), (O$_D$2$py$) | | | | | n | n | n | n | n | n | n | n |

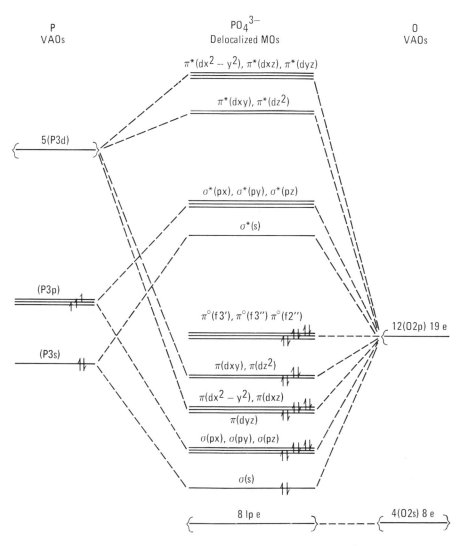

Figure 9-16. Delocalized MO energy level for PO$_4$$^{3-}$.

Thus compared to the P$d_2$ VAO equivalence set, we expect greater interaction of the P$d_3$ VAO equivalence set with the O$p\pi$ VAOs. Note that because of the 45° rotation of the axis system in the tetrahedron in Figure 9-17 compared to Figure 9-7b, the P$d_2$ and P$d_3$ partner sets exchange their $(dxy)$ and $(dx^2 - y^2)$ GO members.

The $\sigma$ MOs in Figure 9-16 can be localized analogously to those in CH$_4$. The bonding $\pi$ MOs, all being $d$-generated, cannot be further localized.

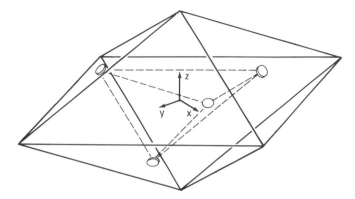

Figure 9-17. The tetrahedral arrangement of four $PO_4^{3-}$ oxygens inscribed in an octahedron.

## 9.5. $VCl_4^-$

Transition metals surrounded by four ligands are either tetrahedral or square planar. Factors influencing the choice of geometry will be discussed later in this section. We begin by considering $VCl_4^-$, which has been determined to be tetrahedral. The delocalized view of this ion can be obtained analogously to that of $PO_4^{3-}$ and the generator table for $VCl_4^-$ given in Table 9-5 is seen to be very similar to Table 9-4. The main difference in these tables is that in contrast to $PO_4^{3-}$, the $(3d)$ VAOs of vanadium lie below the valence $(4s)$ and $(4p)$ VAOs, whereas in $PO_4^{3-}$ the $(3d)$ valence orbitals lie above the $(3s)$ and $(3p)$ orbitals. Consequently the $\pi$ ABMOs fall below the $\sigma$ ABMOs, as shown in Figure 9-18. Thus the additional two electrons in $VCl_4^-$ occupy the $\pi^*(dxy)$ and $\pi^*(dz^2)$ MOs. The presence two unpaired electrons in this anion, as shown in Figure 9-18, has been verified experimentally.

The localized view of $VCl_4^-$ can be obtained analogously to that of $PO_4^{3-}$, except that the two unpaired electrons in $VCl_4^-$ as well as the $\pi$ BMO electrons cannot be localized.

As with octahedral transition metal complexes, a CFT view can also be obtained for other geometries including tetrahedral complexes such as $VCl_4^-$. Again the splittings of $(d)$ VAOs are considered in this model, and because the $(d)$ orbitals are most easily visualized in octahedral symmetry, we again make use of Figure 9-17. Building a model of this figure is the best way of becoming convinced that the $(dxy)$, $(dxz)$, and $(dyz)$ orbitals experience the same environment or field provided by the four negative chloride charges in $VCl_4^-$. Thus these VAOs are degenerate. Moreover, none of these orbitals have the full symmetry of the tetrahedron. That the $(dx^2 - y^2)$ and $(dz^2)$ VAOs also feel identical crystal fields can be appreciated from the observation that neither of these AOs contains the full symmetry of the tetrahedron and hence they form a second partner set. Because the lobes of the triply degenerate set

Table 9-5.  Generator Table for $VCl_4{}^-$ $(26\ e^-)$

| VAO Equivalence Sets | GOs | | | | | | | | | | | |
|---|---|---|---|---|---|---|---|---|---|---|---|---|
| | $s$ | $px$ | $py$ | $pz$ | $dxy$ | $dz^2$ | $dx^2 - y^2$ | $dxz$ | $dyz$ | $f3'$ | $f3''$ | $f2''$ |
| $Vs = (V4s)$ | n | | | | | | | | | | | |
| $Vp = (V4px),\ (V4py),\ (V4pz)$ | | n | n | n | | | | | | | | |
| $Vd_2 = (V3dxy),\ (V3dz^2)$ | | | | | n | n | | | | | | |
| $Vd_3 = (V3dx^2 - y^2),\ (V3dxz),\ (V3dyz)$ | | | | | | | n | n | n | | | |
| $Clp\sigma = (Cl_A 3pz),\ (Cl_B 3pz),\ (Cl_C 3pz),\ (Cl_D 3pz)$ | n | n | n | n | | | | | | | | |
| $Clp\pi = (Cl_A 2px),\ (Cl_A 2py),\ (Cl_B 2px),\ (Cl_B 2py),$ $(Cl_C 2px),\ (Cl_C 2py),\ (Cl_D 2px),\ (Cl_D 2py)$ | | n | n | | n | n | n | n | n | n | n | n |

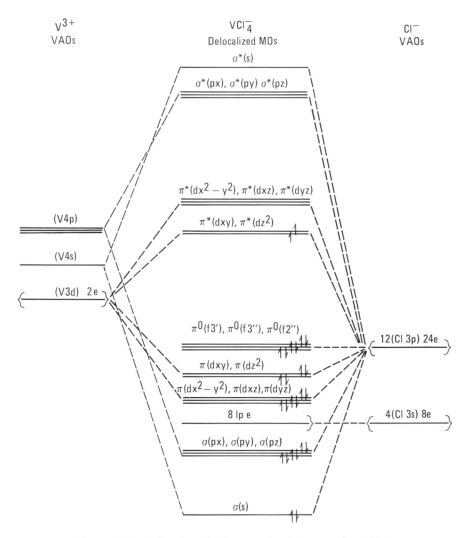

Figure 9-18. Delocalized MO energy level diagram for $VCl_4^-$.

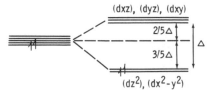

Figure 9-19. Splitting of the metal ($d$) VAOs in a tetrahedral crystal field. The axis system employed here is shown in Figure 9-17.

are closer to the four ligand charges than the lobes of the doubly degenerate set, the triply degenerate orbitals are raised in energy compared to the doubly degenerate orbitals. The splitting pattern for VCl$_4^-$ which results is that shown in Figure 9-19. Notice that the ordering of the ($d$) VAOs in the CFT model is the same as that of the $\pi$ delocalized ABMOs in Figure 9-18, which have mainly vanadium ($3d$) character. The same parallel holds for octahedral CoF$_6^{3-}$ except that the order of the orbitals is inverted and that the doubly degenerate ABMOs for the octahedral case are of the sigma in stead of the pi type. It is interesting to observe here that the differentiation of strong-and weak-field complexes from the measured degree of paramagnetism first occurs at the $d^3$ configuration for tetrahedral coordination complexes whereas it initially occurs at the $d^4$ configuration for octahedral ones. Although admittedly somewhat confusing in the present context, $d^n$ is a standard notation which refers to the orbital occupation number and not to the number of orbitals, as in a hybrid set. Tetrahedral complexes show a remarkable tendency to be of the weak-field type, however, since four ligands are unable to split the ($d$) orbitals as strongly as six.

In closing this section we briefly address the question of why four ligands can frequently surround a transition metal in a square planar array even though there are no lp's above and below the plane as in ICl$_4^-$. A rationale for this observation comes from a comparison of the stabilization energy attained in a tetrahedral and a square planar crystal field and of the relative sizes of the metal ion and the ligand. Most square planar complexes have $d^8$ metal configurations. From Figure 9-20 it can be seen that if the crystal field is large enough to overcome the electron pairing energy, the orbital energy stabilization achieved in the square planar configuration is expected to be

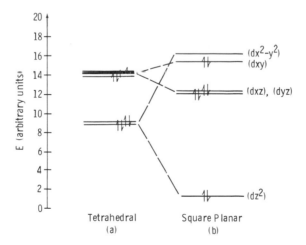

Figure 9-20. Relative energies of metal ($d$) VAOs in a tetrahedral (a) and a square planar (b) crystal field.

greater than in the tetrahedral array (in the absence of countervailing effects, such as large ligand size which would tend to drive the structure to a tetrahedral configuration). It should be pointed out that in general, it has not been possible to predict geometries of new transition metal complexes from simple considerations as unambiguously as geometries can be predicted for compounds of main-group elements, owing to the variety of effects arising from incomplete $(n - 1)d$ orbitals. For this reason, a great deal of challenging research in this area still remains.

The molecular motions of tetrahedral $Z_4$ and $ZY_4$ species can be obtained in the usual way from the appropriate LCAVs represented by the SOs composed of vector-like ($p$) VAOs in Figure 9-8.

# Summary

The delocalized and localized bonding views of $PnH_3$ were seen to be an extension of the considerations employed for generating these views in HChH (Chapter 6). In examining $P_4$ we saw that generating SOs among tangential VAOs in a tetrahedral geometry with the higher GOs $f1'$ and $f1''$ requires either a knowledge of the angle $\theta$ of the conal node, or a prior consideration of the rest of the $f$-generated SOs. After studying the bonding in $CH_4$, in which no unusual features arose, we found that in $PO_4^{3-}$ the unequal splitting for the triply and doubly degenerate BMO–ABMO pairs generated by the $d$ GOs could be rationalized on the basis of geometrical arguments. These considerations were also used in developing the crystal field and ligand field views of $VCl_4^-$. Finally, some of the factors governing the stability of square planar and tetrahedral transition metal complexes were discussed.

Since no fundamentally new procedures for employing GOs were introduced in this chapter, Table 8-10 is still a valid summary of our recipe for the present.

PROBLEMS

1. Sketch the SOs for $PnH_3$ in Table 9-1 and also its molecular motions.

2. Draw the localized MO energy level diagrams for (a) $CH_4$, (b) $PO_4^{3-}$, (c) $VCl_4^-$.

3. Develop the delocalized and localized bonding views for (a) $SbCl_3$, (b) the tetrahedral $C_4$ cluster in the stable tetrahedrane derivative $[CC(CH_3)_3]_4$, (c) $AlCl_4^-$, (d) $TiCl_4$, (e) $PtCl_4^{2-}$ ($d^8$).

4. Generate the molecular motions for (a) $P_4$, (b) $AsH_4^+$.

5. By trigonometric means determine the coefficients of the SOs in Figure 9-8e, l, and s.

6. $Ir_4(CO)_{12}$ consists of a tetrahedral cluster of iridium atoms, each of which is bonded to the carbons of three CO groups. Each member of a set of three CO

groups is colinear with a tetrahedral edge of the tetrahedron and can be considered to donate a pair of electrons from an $sp$ hybrid on the carbon to an $sd^3$ orbital on the metal. Such a set is obtained by hybridizing the $(4s)$ metal VAO with $(3dxy)$, $(3dxz)$, and $(3dyz)$, and the main lobe of each member of this set points to the corners of a tetrahedron. Develop the delocalized view of the cluster binding.

7. CO is a poor Lewis base and $BH_3$ is not a particularly good Lewis acid. The remarkable stability of the adduct $H_3BCO$ has been attributed to back-donation of electron density from B–H $\sigma$ BMOs into the $\pi^*$ MOs of CO. Using GOs between the boron and carbon, sketch the MOs expected from such an interaction and show from an MO energy level diagram why this interaction is expected to lead to increased stability.

8. There is calculational evidence which supports a strong measure of back-donation from filled metal (d) VAOs to $\sigma^*$ P–F MOs in zero-valent metal $PF_3$ complexs with virtually no involvement of (d) VAOs on phosphorus. Using GOs, sketch the MOs expected from the appropriate metal (d) VAOs and the $\sigma^*$ $PF_3$ MOs in octahedral $Cr(PF_3)_6$.

9. From the information given in Table 1 of Appendix IV, calculate the $\theta$ values of the $(d0)$, $(f0)$, $(g0)$, $(g1')$, $(g1'')$, $(g2')$, $(g2'')$, $(h0)$, $(h1')$, $(h1'')$, $(h2')$, $(h2'')$, $(h3')$, and $(h3'')$ AOs. Verify your values where appropriate with those given in Figure 2-6.

10. By drawing the appropriate SOs, show that there are $(3d)$ VAOs on phosphorus that are of the correct symmetry to participate in $\sigma$ bonding in $PO_4^{3-}$.

# CHAPTER 10

# Bipyramidal and Related Molecules

In this chapter we will consider molecules having trigonal and pentagonal bipyramidal symmetry. The tetragonal bipyramid appears to have been skipped here but it should be noted that this geometry is the same as an octahedron if all six vertex atoms are identical. Molecules having trigonal bipyramidal (tbp) symmetry include the cluster anion $B_5H_5^{2-}$ and $PF_5$, depicted in Figure 10-1a and b, respectively. Other examples of trigonal bipyramidal molecules are $PCl_5$(gas), $AsF_5$, and $SbCl_5$. Related to $PF_5$ are $SF_4$ and $BrF_3$ (Figure 10-1c and d) which can be viewed as having, respectively, one and two vertex atoms in the equatorial plane of the tbp replaced by lone pairs. Additional examples of such species are $SeF_4$, $TeCl_4$, $PCl_4$, $SbCl_4$, $ClF_3$, and $XeF_3^+$. Of pentagonal bipyramidal symmetry are $B_7H_7^{2-}$ and $IF_7$ (Figure 10-1e and f). Several metal fluoride complexes also appear to possess this geometry [e.g., $MF_7^{3-}$ (M = Zr, Hf, Nb, U) and $ReF_7$]. Species of higher-order bipyramidal geometries do not appear to have been characterized.

## 10.1. $BrF_3$

This volatile interhalogen compound formed by direct combination of the elements has been determined to have the structure shown in Figure 10-1d. The electron dash structure in Figure 10-1d is similar to the dash structure of $XeF_2$ in Figure 10-1g. We therefore expect the five electron pairs to point toward the corners of a trigonal bipyramid in the localized view. We could have chosen to treat $XeF_2$ as a tbp molecule with three vertex atoms missing, but we have already discussed it as a linear molecule in Chapter 4. It should be recognized from our treatment of $XeF_2$ that the localized description of $BrF_3$ does not necessitate the inclusion of $(d)$ orbitals on the bromine. The

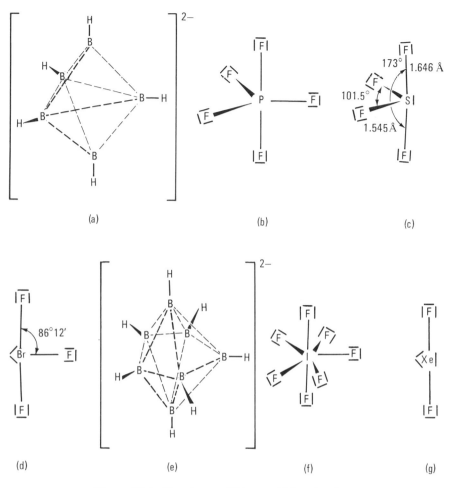

Figure 10-1. Structures of bipyramidal molecules.

electronic structures of $XeF_2$ and $BrF_3$ are quite similar except that one lone pair in the equatorial plane of $XeF_2$ (as shown in Figure 10-1g) is replaced by a bond pair in a Br—F link (Figure 10-1d). Recall also that it is a general rule of VSEPR theory that in the localized view, five electron pairs arrange themselves in a trigonal bipyramidal array with lone pairs (if any) occupying equatorial positions.

For convenience, we will choose the coordinates shown in Figure 10-2 (which, for simplicity, do not include the slight bending of the vertical molecular axis away from 180°). From the symmetry of the molecule it is clear that the fluorine sites are not equivalent. This site inequivalence also induces chemical inequivalence in these atoms since their valence electronic environments are different. Thus $F_A$ and $F_B$ are equivalent to each other but inequivalent with respect to $F_C$. We assign lone pairs to the $(4s)$ and $(4px)$ VAOs on

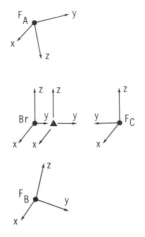

Figure 10-2. Axis system for $BrF_3$.

Table 10-1. Generator Table for $BrF_3$ (6 e$^-$)

|  | GOs | |
| --- | --- | --- |
| VAO Equivalence Sets | $s$ | $pz$ |
| $Brpy = (Br4py)$ | n | |
| $Brpz = (Br4pz)$ | | n |
| $Fpz = (F_A2pz), (F_B2pz)$ | n | n |
| $Fpy = (F_C2py)$ | n | |

bromine, the $(2s)$, $(2px)$, and $(2pz)$ VAOs on $F_C$, and the $(2s)$, $(2px)$, and $(2py)$ VAOs on $F_A$ and $F_B$. [The $(Br4s)$ lone pair must actually possess some $(p)$ character to be stereoactive.] From the remaining VAOs we construct Generator Table 10-1.

Since there are three electron pairs for the five MOs of this molecule, two MOs will be unoccupied. These two will obviously be the $s$- and $pz$- generated ABMOs. Thus we conclude that two of the three $s$-generated MOs are occupied. We now establish that the SO $\tilde{\sigma}(s|Fpz)$ in Figure 10-3a can be taken to be an occupied NBMO by exploiting the fact that any two orbitals formed as orthogonal linear combinations of the occupied canonical MOs can serve as filled MOs generated by the same **GO**. As explained earlier, we can always choose an orbital pair so that the third SO will not contribute. If we choose the orbital pair so that $\tilde{\sigma}(s|Brpy)$ (Figure 10-4a) is the excluded orbital, then the MO which does not contain it is a linear combination of $\tilde{\sigma}(s|Fpz)$ and $\tilde{\sigma}(s|Fpy)$ (Figure 10-4b and c) and hence is manifestly nonbonding, since these SOs are too far separated to overlap effectively. On the other hand, if we choose the orbital pair so that $\tilde{\sigma}(s|Fpy)$ is the excluded orbital, then the MO which does not contain it is a linear combination of $\tilde{\sigma}(s|Fpz)$ and $\tilde{\sigma}(s|Brpy)$ and hence is essentially nonbonding because these SOs are effectively orthogonal. Now,

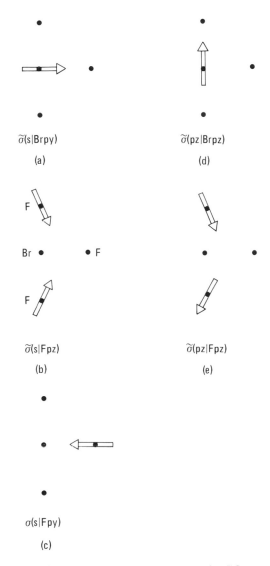

Figure 10-3. SOs of BrF$_3$ as given in Table 10-1. The **GO**s are not shown.

since the two NBMOs must be essentially the same orbital, we conclude that the NBMO is basically $\tilde{\sigma}(s|\mathrm{F}pz)$, which is contained in both. Another way of drawing the same conclusion is to observe that linear combination of $\tilde{\sigma}(s|\mathrm{Br}py)$ and $\tilde{\sigma}(s|\mathrm{F}py)$ obviously leads to a BMO and an ABMO, allowing $\tilde{\sigma}(s|\mathrm{F}pz)$ to serve as a manifestly NBMO.

From the preceding discussion it should be noted that the atom contributions to the various MOs break down the orbital structure into a triatomic and a diatomic part. Thus $\sigma^0(s)$, $\sigma(pz)$, and $\sigma^*(pz)$ contain contributions from

$F_A$, $F_B$, and Br, while $\sigma(s)$ and $\sigma^*(s)$ involve contributions from Br and $F_C$. A BMO, an NBMO, and an ABMO are associated with the triatomic moiety while a BMO and an ABMO describe the diatomic portion, and the two bonding patterns operate independently in the two parts of the molecule. Of the twenty-eight valence electrons in the molecule, only six have not been assigned as lp's, and they occupy the MOs $\sigma(s)$, $\sigma^0(s)$, and $\sigma(pz)$. The energy level diagram in Figure 10-4 shows the parallel between $BrF_3$ and $XeF_2$ which was made at the beginning of this section. Thus there are only two bonding electrons delocalized over three centers ($F_A$, Br, and $F_B$), as was the case in $XeF_2$. From this figure it can be rationalized why the $BrF_A$ and $BrF_B$ bond lengths are equal but longer than $BrF_C$. The former are 3-center 2-electron

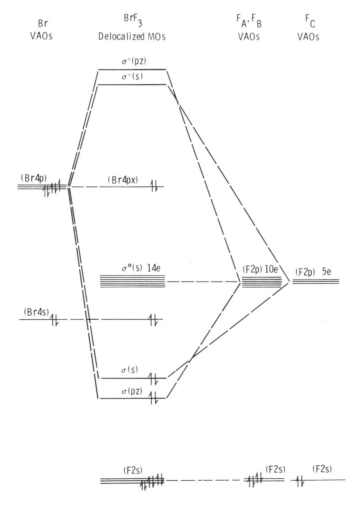

Figure 10-4. Delocalized MO energy level diagram for $BrF_3$.

bonds (bond order = 0.5) while the latter is a 2-center 2-electron bond (bond order = 1.0).

It is easily recognized from the delocalized bond picture that the bonding in the diatomic part is already localized. The only further localization required is that which can be carried out on the triatomic part and the lone pairs. The delocalized $\sigma$ bonding system along the $F_A$—Br—$F_B$ axis can be localized into two half-bonds as was the case for XeF$_2$. The lp's on bromine can be localized into ($spx$) VAO hybrids and those on the fluorines can be localized into ($sp^2$) VAO hybrids. The localized MO diagram is shown in Figure 10-5.

Although the generation of molecular motions of BrF$_3$ follows our usual

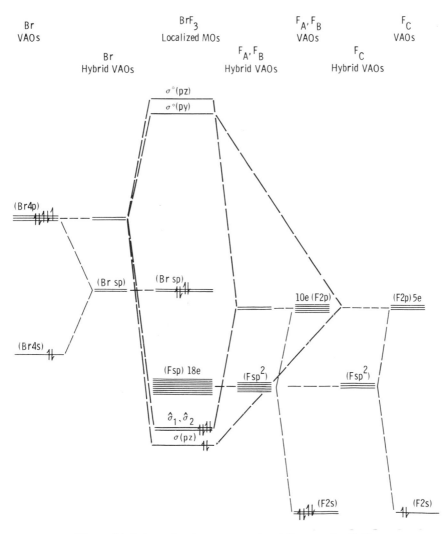

Figure 10-5. Localized MO energy level diagram for BrF$_3$.

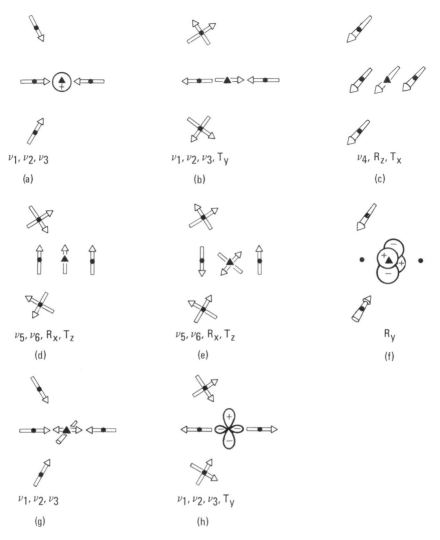

Figure 10-6. SMs for T-shaped $ZY_3$, as given in Table 10-2.

recipe, a somewhat unusual feature emerges because of the molecule's low symmetry. From the **GO**-generated SMs drawn in Figure 10-6 and summarized in Generator Table 10-2, we see that the $s$ **GO** calls in SMs from three AV equivalence sets. Interestingly, $py$ generates the same set of SMs plus an additional one. (Figure 10-6b). Thus the column entries headed by the less inclusive $s$ **GO** in Table 10-2 is enclosed in parentheses and can be disregarded. Similarly $dx^2 - y^2$ (Figure 10-6g) and $dz^2$ (Figure 10-6h) generate the same SMs as $s$ and $py$, respectively, and can be ignored. It is clear from Figure 10-6b that the translation $T_y$ must arise from an AV equivalence set combina-

Table 10-2. Generator Table for T-Shaped ZY$_3$

| AV Equivalence Set$^a$ | GOs | | | | | | | |
|---|---|---|---|---|---|---|---|---|
| | $s$ | $px$ | $py$ | $pz$ | $dz^2$ | $dx^2 - y^2$ | $dxz$ | $dyz$ |
| Br$x$ | | m | | | | | | (m) |
| Br$y$ | (m) | | m | (m) | (m) | | | |
| Br$z$ | | | | m | | | | |
| F$x$a | | m | | | | | m | (m) |
| F$y$a | | | m | m | (m) | | | |
| F$z$a | (m) | | m | m | (m) | (m) | | |
| F$x$e | | m | | | | | | (m) |
| F$y$e | (m) | | m | | (m) | (m) | | |
| F$z$e | | | | m | | | | |

$^a$Here "a" denotes the axial Y atoms and "e" denotes the equatorial Y atom.

tion involving F$y$a + Br$y$ − F$y$e + F$z$a. The **GO** $pz$ calls in the same AV equivalence sets (Figure 10-6d) as $dyz$ (Figure 10-6e). $T_z$ and $R_x$ in Figure 10-6d clearly arise from linear combinations of the type Br$z$ + F$z$e − F$z$a + F$y$a and Br$z$ − F$z$e + F$y$a − F$z$a, respectively. As expected from the $3n - 6$ rule, six vibrational modes are generated.

## 10.2. SF$_4$

Sulfur tetrafluoride is a gas formed in the oxidative fluorination of SCl$_2$. In its electron dash structure in Figure 10-1c sulfur is presumed to be hypervalent since it has an expanded octet involved in $\sigma$ bonds and lp's. This electron dash structure permits the formal charges on the atoms to be zero. Our experience with isoelectronic BrF$_3$ and XeF$_2$ suggests that in the localized view of SF$_4$ one lp will be in the equatorial plane of a tbp electron pair arrangement. The experimentally determined structure in Figure 10-1c parallels that of BrF$_3$, with the longer bonds bent away from 180° because of lp–bp repulsions. The large electronegativity difference between sulfur and fluorine gives rise to strong ionicity in the S–F bonds.

Using the axis system in Figure 10-7, in which we ignore the slight axial angular distortion, the electron dash structure in Figure 10-1c indicates that we can assign the F$_A$, F$_B$ lone pairs to the (F2$s$), (F2$px$), and (F2$py$) VAOs and the F$_C$, F$_D$ lone pairs to the (F2$s$), (F2$px$), and (F2$pz$) VAOs. The sulfur lp is assigned to the (S3$s$) VAO. In constructing Generator Table 10-3 with the aid of the pictorial representations of the SOs in Figure 10-8, similarities to the generator table and SOs for BrF$_3$ are revealed. Thus there are three SOs generated by $s$. Like BrF$_3$ (and several other molecules which we have considered), SF$_4$ is an example of a system for which two SOs interact mainly with each other but not (significantly) with a third. Hence we expect an $s$-generated BMO–NBMO–ABMO set of MOs, with the NBMO consisting mainly of the F$pz$ SO because of its poor overlap with S$py$ and its zero overlap

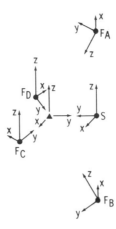

Figure 10-7.  Axis system for $SF_4$.

Table 10-3.  Generator Table for $SF_4$ (8 e$^-$)

| VAO Equivalence Sets | GOs | | |
| --- | --- | --- | --- |
|  | $s$ | $px$ | $pz$ |
| $Spx = (S3px)$ |  | n |  |
| $Spy = (S3py)$ | n |  |  |
| $Spz = (S3pz)$ |  |  | n |
| $Fpz = (F_A 2pz), (F_B 2pz)$ | n |  | n |
| $Fpy = (F_C 2py), (F_D 2py)$ | n | n |  |

with $Fpy$. The bonding and antibonding linear combinations are then seen to arise from the $(S3py)$ VAO and the $Fpy$ SO. A second BMO–ABMO pair for $SF_4$ results from linearly combining $\tilde{\sigma}(px|Fpy)$ and $(S3px)$ (Figure 10-8d, e). The BMO–ABMO pair which stems from linear combinations of the $pz$-generated SOs (Figure 10-8f, g) is analogous to that formulated earlier for $BrF_3$. Thus we see that $SF_4$ has two triatomic parts: a bent one in the equatorial plane and a linear one along the $z$ axis. Drawing a qualitative energy level diagram is straightforward and is left as an exercise. All three of the BMOs are filled, giving average bond orders of 1.0 and 0.5 for the shorter equatorial bonds and the longer axial bonds, respectively.

Localization of the occupied MOs associated with the axial triatomic part of the molecule is analogous to that in $BrF_3$ and $XeF_2$ and hence will not be discussed further. Localization of the occupied MOs of the equatorial triatomic moiety also poses no new considerations since this can be accomplished in an analogous manner to the localization of the occupied $\sigma$ bonding MOs of $H_2O$. In the present case, the bond angle in the equatorial plane is 101.5°.

Generating the molecular motions of a species having the geometry of $SF_4$ is done in the usual way and this is left as an exercise.

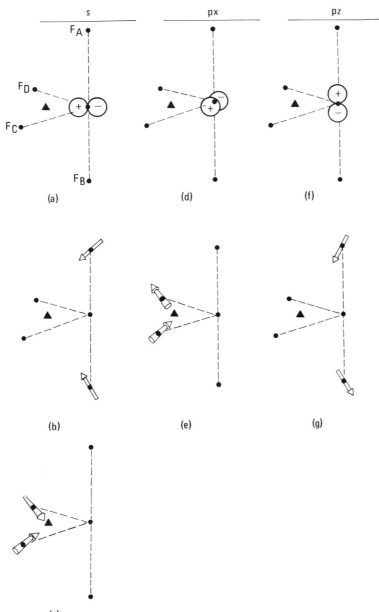

Figure 10-8. SOs for SF$_4$ as given in Table 10-3. The **GO**s are not shown in the drawings.

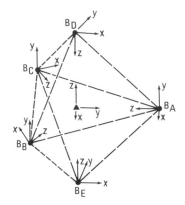

Figure 10-9. Axis system for $B_5H_5{}^{2-}$.

Table 10-4. Generator Table for $B_5$ Cluster in $B_5H_5{}^{2-}$ (12 e$^-$)

| VAO Equivalence Sets | GOs | | | | | | |
|---|---|---|---|---|---|---|---|
| | $s$ | $px$ | $py$ | $pz$ | $dyz$ | $dxz$ | $f3''$ |
| B$spa$-in = (B$_D spi$), (B$_E spi$) | n | | | n | | | |
| B$pxa$ = (B$_D 2px$), (B$_E 2px$) | | n | | | n | | |
| B$pya$ = (B$_D 2py$), (B$_E 2py$) | | | n | | | n | |
| B$spe$-in = (B$_A spi$), (B$_B spi$), (B$_C spi$) | b | a | a | | | | |
| B$pxe$ = (B$_A 2px$), (B$_B 2px$), (B$_C 2px$) | | b | b | | | | a |
| B$pye$ = (B$_A 2py$), (B$_B 2py$), (B$_C 2py$) | | | | b | a | a | |

## 10.3. $B_5H_5{}^{2-}$

This anion is another example of an electron-deficient system. Anions of the type $B_nH_n{}^{2-}$ where $n = 6$–$12$ are well known and all have been found to possess *closo* (i.e., closed) configurations. Thus, for example, the geometries of $B_6H_6{}^{2-}$, $B_7H_7{}^{2-}$, and $B_{12}H_{12}{}^{2-}$ are octahedral, pentagonal bipyramidal, and icosahedral, respectively. It is not unreasonable to expect that the as yet uncharacterized $B_5H_5{}^{2-}$ ion will have tbp symmetry (Figure 10-1a) since $B_3C_2H_5$, an isostructural isoelectronic analogue, is known.

The B—H links in $B_5H_5{}^{2-}$ are 2-center 2-electron bonds and we reserve the appropriate (B$sp^\alpha$) hybrid on each boron for bonding to hydrogen. Note that $\alpha$ will undoubtedly be different for the apical and equatorial boron VAOs. As can be seen in the axis system of Figure 10-9, the inwardly directed orthogonal hybrid VAOs divide into two sets, the apical hybrids (B$spa$-in) and the equatorial hybrids (B$spe$-in). Furthermore, the two (B$2px$) and (B$2py$) VAOs, which are directed tangentially around the trigonal bipyramidal framework, also form apical and equatorial VAO sets. Having accounted for ten electrons in localized B—H bonds, there are twelve electrons left to bind the boron cluster.

Using the axis system shown in Figure 10-9, we can construct Generator Table 10-4 with the help of Figure 10-10. The **GOs** $px$ and $py$ in Table 10-4 are

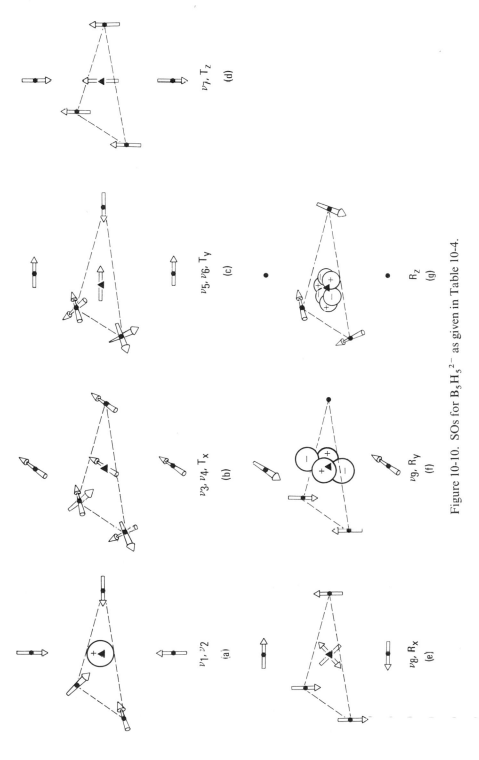

Figure 10-10. SOs for B$_5$H$_5$$^{2-}$ as given in Table 10-4.

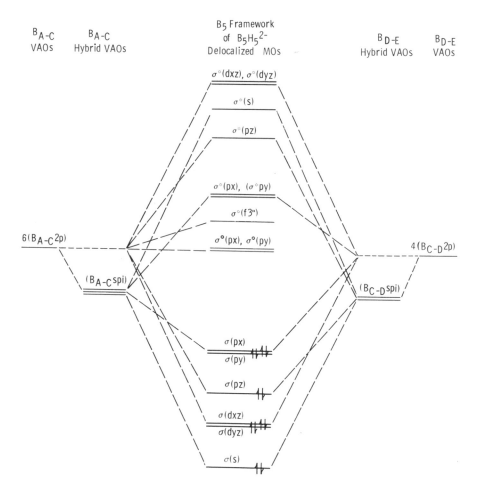

Figure 10-11. Delocalized MO energy level diagram for $B_5$ framework of $B_5H_5^{2-}$.

seen to each generate three SOs. By linearly combining the SOs generated by the $s$, $px$, $py$, and $pz$ **GO**s, four BMOs can be obtained which will house eight of the twelve electrons which bind the borons together. Linear combinations of the antibonding and nonbonding SOs generated by $dxz$ and $dyz$ **GO**s give rise to two BMO–ABMO pairs. The two BMOs are occupied by the remaining four electrons. From the energy level diagram depicted in Figure 10-11 it is seen that the overall B—B bond order is 2/3 since there are nine edges on a trigonal bipyramid.

Although it is clear from Table 10-4 that there should be a total of six BMO–ABMO pairs and three NBMOs, the ordering among the BMOs and the ABMOs cannot be predicted on a qualitative basis and no order should be inferred from the spacings shown in the energy level diagram in Figure 10-11.

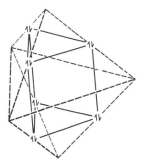

Figure 10-12. Localized bond pair positions indicated by the $sp^3dxzdyz$ GOs.

It is seen from this diagram, however, that the twelve electrons available for boron cluster binding fill all of the BMOs but none of the NBMOs.

Hybridization of the **GO** set associated with the BMOs of Figure 10-11 leads to an $sp^3d^2$ hybrid set. When such a set contains a $dxz$ and a $dyz$ orbital, the main lobes of its members point to the vertices of a trigonal prism. In Figure 10-12 is shown how a trigonal prism can be inscribed in a trigonal bipyramid. Recall that in the previous chapter's discussion of the localized view of $P_4$ we saw that if an $sp^3d^2$ hybrid set contains a $dxy$ and $dz^2$ orbital, then the main lobes of the set point to the vertices of an octahedron. Thus the directionality of the hybrids in Figure 10-12 indicates that the six **BMO** electron pairs should be localized at the edges of the tbp, as indicated in the figure.

The question regarding the choice of boron VAO orientation to accommodate our localization scheme can be resolved by considering the apical boron VAOs to consist of a set of approximately $(sp^3)$ hybrids, one of which is used to make a 2-center 2-electron bond with a hydrogen, while the other three are directed over the three edges of the trigonal bipyramid as shown in Figure 10-13. In order to avoid bent bonds with the equatorial borons, we may wish to place more $(p)$ character in the edge-oriented lobes of the apical boron hybrids, which would bring them more parallel to the edges of the trigonal bipyramid. Of course more $(s)$ character must then be moved into the hybrids pointed toward the apical hydrogens. By assuming the VAOs on the equatorial borons to consist of $(sp^2)$ orbitals, as schematically depicted in Figure 10-13 (wherein the outwardly directed hybrids are each used to make a 2-center 2-electron bond with an equatorial hydrogen), we can make 2-center 2-electron bonds between pairs of equatorial and apical boron hybrids as shown in the localized energy level diagram in Figure 10-14. In this energy level diagram the delocalized ABMOs shown in Figure 10-11 are omitted since we have changed the hybridization on the borons and thus it would not be straightforward to depict in Figure 10-14 the dashed-line connections to the ABMOs in Figure 10-11. Recall also that the localization of ABMOs is not very meaningful anyway.

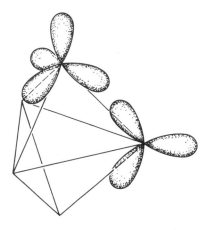

Figure 10-13. Sketch showing directions of apical $sp^3$ and equatorial $sp^2$ VAOs conveniently oriented for localized 2-center bonds.

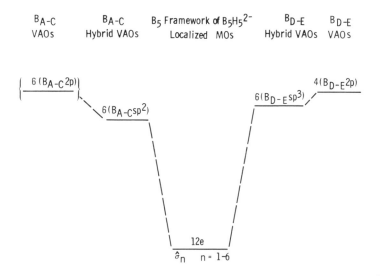

Figure 10-14. Localized MO energy level diagram for $B_5$ framework of $B_5H_5^{2-}$. The delocalized NBMOs and ABMOs are not shown.

It should be noted that the orientation of the $sp^2$ VAO hybrids on the equatorial borons requires that there be three boron $(2px)$ VAOs (one from each boron) oriented tangentially around the equatorial plane. These $(2px)$ VAOs form the delocalized NBMOs and the $f3''$-generated ABMO of the molecule. Thus the localized bond picture we have developed permits electron pair binding between equatorial and apical pairs of borons, but not between equatorial borons even though they would presumably be within bonding distance of one another.

Table 10-5.  Generator Table for PF$_5$ (10 e$^-$)

| VAO Equivalence Sets | GOs | | | | |
|---|---|---|---|---|---|
| | $s$ | $dz^2$ | $px$ | $py$ | $pz$ |
| P$s$ = (P3$s$) | n | (n) | | | |
| P$p$a = (P3$pz$) | | | | | n |
| P$p$e = (P3$px$), (P3$py$) | | | n | n | |
| F$pz$a = (F$_D$2$pz$), (F$_E$2$pz$) | n | (n) | | | n |
| F$pz$e = (F$_A$2$pz$), (F$_B$2$pz$), (F$_C$2$pz$) | n | (n) | n | n | |

# 10.4.  PF$_5$

Fluorination of phosphorus produces PF$_5$, a colorless, stable gas. The electron dash structure (Figure 10-1b) is similar to the geometrical arrangement of the five electrons pairs around the central atom in XeF$_2$, BrF$_3$, and SF$_4$. The expected trigonal bipyramidal structure of this hypervalent molecule has been confirmed by electron diffraction experiments; the axial bond lengths are slightly longer than the equatorial ones. Both types of bonds contain strong ionic character.

After assigning the lp's to fluorine (2$s$), (2$px$), and (2$py$) VAOs, Generator Table 10-5 can be constructed using the coordinate system in Figure 10-9. This table reflects a fact we later use, namely, that $s$ and $dz^2$ generate the same SOs. As in BrF$_3$ and SF$_4$, the apical fluorines and the equatorial fluorines in PF$_5$ interact with the central atom but do not interact with each other. Since PF$_5$ has ten valence electrons and a total of nine MOs according to Table 10-5, we expect that one ABMO corresponding to each of the GOs will be unoccupied and that the $s$-generated NBMO will be occupied. The linear combination of $\tilde{\sigma}(s|$F$pz$e$)$ and $\tilde{\sigma}(s|$F$pz$a$)$ is manifestly nonbonding. In addition the BMOs of the BMO–ABMO pairs generated by $px$, $py$, and $pz$ are occupied. The BMO $\sigma(pz)$ forms a 3-center bond in the linear triatomic part of PF$_5$, $\sigma(px)$ and $\sigma(pz)$ form 4-center bonds in the tetra-atomic equatorial plane, and $\sigma(s)$ which is a combination of all three SOs, dominated by $\tilde{\sigma}(s|$P$s)$ forms a 6-center bond. Ascribing a bond order of approximately 1/5 in each P—F link to the contribution from $\sigma(s)$, we then have a bond order of $2/3 + 1/5 = 13/15$ in each of the P—F equatorial links. Similarly, we have $1/2 + 1/5 = 0.7$ bond order in each P—F axial link. Drawing the delocalized MO energy level diagram is left as an exercise.

In establishing the localized bond picture we are faced with the problem that two occupied MOs are generated by the same GO. This situation, which we have encountered previously in our discussion of N$_2$, sometimes occurs for electron-rich systems where NBMOs and/or ABMOs are occupied. If we choose to localize only the occupied BMOs, we must deal with an $sp^3$ GO set. To inscribe the tetrahedral symmetry of such a GO set inside the trigonal bipyramidal PF$_5$ is awkward at best. If we include the GO of the occupied NBMO (namely, $s$) we face the strange prospect of forming hybrid GOs with the $sspxpypz$ set which includes two $s$ GOs. We can circumvent this problem by recalling that the $s$ **GO** is not the only generator of the first column in Table 10-5. In fact, any **GO** with axial symmetry about the $z$ axis which is also symmetric on reflection through the $xy$ plane gives rise to the same SOs. In addition to $s$, this set includes $dz^2$, $g0$, and still higher **GO**s, all of which are characterized by having either zero conal nodes in the case of $s$, or an even number of conal nodes (two for $dz^2$, four for $g0$, etc.). On the basis of the nodal structure of the orbitals, it is in fact easily seen that the BMO and the NBMO are most naturally associated with the $s$ **GO** and the $dz^2$ **GO**, respectively. With this assignment of the **GO**s to the MOs, the localized isoenergetic MOs are formed from a set of $sp^3d$ **GO**s. The orbitals of this set are not all spatially equivalent but consist of two equivalent **GO**s directed oppositely along the $z$ axis and three equivalent **GO**s in the equatorial plane. Having done this, we must realize that by introducing the $dz^2$ **GO**, we have tacitly assumed that a corresponding $3dz^2$ VAO is generated on phosphorus; a VAO we did not include in the delocalized view! It is therefore not surprising that the bond order in our localized picture is 1.0 instead of the fractional bond orders we saw in the delocalized view.

By creating an isoenergetic set of $sp^3d$ **GO**s, we find that the five localized MOs $\sigma_n$ ($n = 1 - 5$) are also at the same energy. Because the axial and equatorial bond lengths of $PF_5$ are different, however, we might expect that their localized bond energies are also slightly different. We can incorporate this difference into our localized view by creating an $spxpy$ **GO** set for the bonds in the equatorial plane and a $pzdz^2$ **GO** set for the bonds along the three-fold axis. This provides us with three isoenergetic localized equatorial MOs at a somewhat lower energy than the two isoenergetic localized axial MOs.

The molecular motions of $Z_5$ and $ZY_5$ species having tbp symmetry are obtained by our usual route, as shown in Figure 10-10 for $Z_5$. Generating the motions of tbp $ZY_5$ is left as an exercise.

## 10.5. $IF_7$

Heating fluorine with iodine gives $IF_7$. The electron dash structure shown in Figure 10-1f indicates that there are seven bond pairs around iodine. From VSEPR considerations, these seven bond pairs are predicted to arrange themselves into a pentagonal bipyramidal geometry and this has been experimentally verified for $IF_7$. Since the equatorial bond pairs are spanned by smaller angles (72°) than the 90° angles separating the axial and equatorial bond pairs, the axial I—F links are expected from VSEPR considerations to be shorter than the equatorial I—F bonds. This has also been verified experimentally for these polar linkages.

After assigning lone pairs to the fluorine $(2s)$, $(2px)$, and $(2py)$ VAOs, we construct Generator Table 10-6 with the aid of the axis system in Figure 10-15.

Table 10-6. Generator Table for IF$_7$ (14 e$^-$)

| VAO Equivalence Sets | GOs | | | | | |
| --- | --- | --- | --- | --- | --- | --- |
| | $s$ | $px$ | $py$ | $pz$ | $dxy$ | $dx^2 - y^2$ |
| I$s$ = (I5$s$) | n | | | | | |
| I$pe$ = (I5$px$), (I5$py$) | | n | n | | | |
| I$pa$ = (I5$pz$) | | | | n | | |
| F$pe$ = (F$_A$2$pz$), (F$_B$2$pz$), (F$_C$2$pz$) | n | n | n | | n | n |
| (F$_D$2$pz$), (F$_E$2$pz$) | | | | | | |
| F$pa$ = (F$_F$2$pz$), (F$_G$2$pz$) | n | | | n | | |

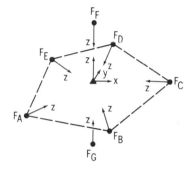

Figure 10-15. Axis system for IF$_7$.

Like PF$_5$, IF$_7$ is an example of a system in which the apical fluorines and the equatorial fluorines interact with the central atom to slightly different extents, and do not interact with each other. Thus, as before, we seek a BMO–NBMO–ABMO set from the $s$-generated MOs in Table 10-6. Further examination of this table tells us that in addition to the BMO and the NBMO arising from the $s$-generated SOs, we have three BMOs which stem from the linear combination of the pairs of $p$-generated SOs. The SOs generated by $dxy$- and $dx^2 - y^2$-generated SOs are both NBMOs, since the equatorial fluorines are not close enough to each other to interact. The seven electron pairs available for bonding are therefore housed in the four BMOs and the three NBMOs. The bond order along the axis of IF$_7$ is 1/2 plus 1/7, and that in the equatorial plane is 2/5 plus 1/7.

In generating a localized view of IF$_7$ we face a problem similar to that encountered in PF$_5$, namely, that two of the seven occupied delocalized MOs of IF$_7$ are generated by $s$. Since $dz^2$ generates the same MOs as $s$, we conclude that the **GO** hybrid set which would produce seven energy-equivalent localized MOs for these seven electron pairs is $sp^3d^3$. From the geometry of IF$_7$ we see that all the members of this hybrid set are not spatially equivalent. (Recall that a similar inequivalence of the orbitals occurred in the $sp^3d$ set for PF$_5$.) The two oppositely directed $pd$ **GO** hybrids along the axis and the five $sp^2d^2$ **GO**s arranged in a pentagonal array in the equatorial plane lead to the electron

Table 10-7. A Useful Summary of Procedures for Generating Pictorial Representations of Delocalized and Localized Bonding Views of a Molecule and for Visualizing its Vibrational Modes

| Delocalized View | Localized View | Normal Modes |
|---|---|---|
| 1. Decide on best electron dash structure using lowest possible formal charges and drawing resonance forms. In such forms, some lp's may be involved in a $\pi$ system. | 1. Decide on which set(s) of doubly occupied delocalized MOs to localize. | 1. Identify the motional atomic vector (AV) equivalence sets and list them in the generator table. |
| 2. Establish the molecular geometry from VSEPR theory if possible. For transition metal complexes and boron hydrides, other considerations apply. | 2. Hybridize the **GO**s which give rise to the sets(s) in 1.<br><br>3. Using the hybridized **GO**s, sketch the SOs. *Treat groups of inequivalent peripheral atoms separately.* | 2. Using **GO**s, sketch the symmetry motions (SMs).<br><br>3. Fill in the rest of the generator table. (i.e., **GO**s and m entries). |
| 3. After preassigning lp's and localized bp's to ($s$) and suitably oriented ($p$) VAOs, identify all the VAO equivalence sets (hybrids may also be used) and list them in the generator table. | 4. Sketch the MOs, being aware of the consequences of a central atom on which the full set of SOs (hybridized VAOs) is not generated by the hybridized **GO** set. If necessary, re-hybridize peripheral atom VAOs and rotate hybrid **GO**s to locate bp's between atoms. | 4. Take appropriate linear combinations of SMs generated by the same **GO** and sketch all the molecular motions.<br><br>5. Identify the normal vibrational modes. |
| 4. Using **GO**s located at a molecular center or centroid, sketch all the SOs, recalling that #SOs = #VAOs. *Treat groups of inequivalent peripheral atoms separately.* | 5. Draw the resulting MO energy level diagram and occupy the appropriate levels with electrons.<br><br>6. Verify that the bond order per link is the same as in the delocalized view. | |
| 5. Fill in rest of generator table (i.e., the necessary **GO**s and the $n$, $a$ or $b$ entries). Establish partner **GO** and SO sets. | | |
| 6. Take linear combinations of SOs generated by the same **GO** and sketch all the MOs. | | |
| 7. Draw an MO energy level diagram and occupy the appropriate levels with electrons. | | |
| 8. Deduce the bond order per link. | | |

dash structure for $IF_7$ shown in Figure 10-1f. As with $PF_5$, we recognize from the delocalized view of $IF_7$ that the bond order of the localized bonds we have created is not 1.0. This higher bond order is a consequence of our localized view in which we tacitly assumed that the ($I5dz^2$) VAO is included. We thus conclude that in cases such as $PF_5$ and $IF_7$, wherein a **GO** appears more than once in the occupied delocalized MO set we wish to localize, we should either refrain from localizing them, or we should be prepared to introduce a new central atom VAO.

## Summary

The lower symmetry of all of the molecules examined in this chapter allowed us to approach their delocalized bonding views by breaking the molecule into axial and equatorial components. In our discussions of $BrF_3$ and $SF_4$ we learned how to extract the essentially NBMO from a set of three SOs generated by the same **GO**. In generating the localized views of $PF_5$ and $IF_7$ we learned that caution must be exercised when different members of occupied delocalized MO sets are generated by the same **GO**.

In Table 10-7 is an amended summary of our procedures with **GO**s.

### Problems

1. Develop the localized view of $SF_4$.

2. Draw delocalized MO energy level diagrams for (a) $PF_5$ and (b) $IF_7$.

3. Sketch the molecular motions for (a) $XeF_3^+$, (b) $SbCl_5$, and (c) $SF_4$.

4. From a consideration of the delocalized MO energy level diagram for $B_5H_5^{2-}$ (Figure 10-11) comment on the feasibility of reducing this ion to the as yet unrealized $B_5H_5^{4-}$. Describe its expected magnetism.

5. By irradiating a mixture of $NF_3$ and fluorine at $-200°C$, a substance is formed for which the late R.E. Rundle found some evidence supporting the formation of $NF_5$. Formulate the delocalized and the localized view of this molecule.

6. Give the electron structure of $F_3ClO_2$, remembering to make the formal charges as low as possible. Sketch the structure of the molecule. (Hint: Both oxygens are in identical geometrical environments.) Justify your structure on the basis of VSEPR arguments and the relative sizes of the peripheral atoms.

7. Develop the delocalized bonding view of $Sn_5^{2-}$.

8. Develop the delocalized and localized bonding views of $IF_7$ in which iodine ($d$) VAOs are included. The axial and equatorial bonds in these molecules are not isoenergetic.

9. Develop the delocalized view of $B_3C_2H_5$. The carbons lie on a three-fold axis of symmetry.

# CHAPTER 11
# Prismatic Molecules

Prismatic species without central atoms are rare. The $Te_6^{4+}$ ion is an example of a trigonal prism and prismane (Figure 11-1a) is the as yet unknown parent compound of several characterized derivatives which have been found to contain the trigonal prismatic array of carbon atoms. Cubane (Figure 11-1b) has been synthesized by photolytically dimerizing cyclobutadiene that has been stabilized in a metal complex.

Examples of prismatic molecules containing central atoms are arene organometallic "sandwich" compounds of which many are known. The arene ligands are typically cyclopentadienide, benzene, or cyclooctatetraenide, and two of these planar moieties "sandwich" a metal atom between them. In Figure 11-2 are shown two conformations of *bis*-cyclopentadienide metal complexes which have been experimentally verified. There is also structural evidence that the hydrogens on the $C_5$ rings are bent slightly away from the molecular center and out of the carbon plane. Since we are concerned with prismatic rather than antiprismatic molecules in this chapter, we will restrict our attention to the pentagonal prismatic conformation in Figure 11-2a. *Bis*-arene complexes are known for a variety of transition metals [including lanthanides and actinides in the case of $[(\eta^8\text{-}C_8H_8)_2M]^x$ and metal oxidation states from $+1$ to $+4$ can be stabilized in such organometallic compounds. The $\eta^n$ symbol denotes that $n$ ring atoms of the ligand are bonded to the central atom. It should be recognized that the full set of such ring atoms do not always bond to a metal atom, in which case the complex is no longer a true "sandwich" compound. Cubic ions such as $U(NCS)_8^{4-}$, $UF_8^{3-}$, and $NpF_8^{3-}$ are also known.

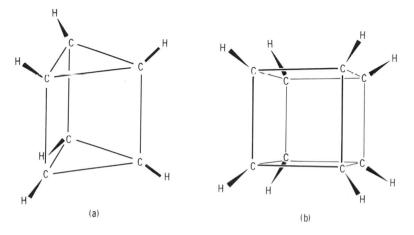

Figure 11-1. Prismane (a) and cubane (b).

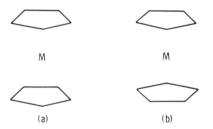

Figure 11-2. Eclipsed (a) and staggered (b) conformations of *bis*-cyclopentadienide metal complexes.

## 11.1. C$_6$H$_6$ (Prismane)

The electron dash structure for prismane is shown in Figure 11-1a. Since there are thirty valence electrons for a total of the fifteen C—C and C—H links, this carbon cluster system is not electron deficient. The C—C bonds are strained, however, because of the non-tetrahedral carbon angles. This as yet unknown molecule is therefore prone to bond-breaking reactions and is unstable with respect to thermal rearrangement to benzene, which is un-strained and is stabilized by resonance.

We first localize the prismane C—H bond pairs in carbon ($sp^\alpha$) VAO hybrids. The VAOs on a given carbon atom which are used to bond the carbon framework must be taken to be orthogonal to $sp^\alpha$. In the spirit of previous examples, we could pick an orthogonal ($sp$) hybrid oppositely directed from ($sp^\alpha$) and toward the molecular center indicated by the triangle in Figure

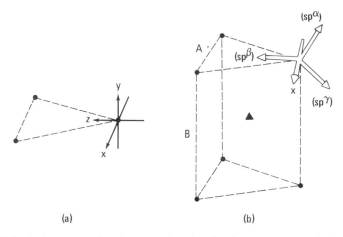

Figure 11-3. Axis system for the top triangle of prismane (a). In (b) is shown a convenient configuration of VAOs on one of the carbons of prismane.

11-3b. The remaining two $(2p)$ VAOs could then lie at 90° to one another in a plane perpendicular to the two hybrids. However, because we can make use of what we learned about the bonding in triangular species in Chapter 5, it is convenient to pick one VAO to lie in the plane of the triangle (see Figure 11-3b), just as we did in the case of $N_3^+$. The remaining VAO, $sp^\gamma$, is now determined. It lies more or less (but not exactly) along the C—C link connecting the lower and the upper triangle. It is slanted slightly with respect to the vertical C—C bonds toward the outside of the trigonal prism at an angle that depends on the orientation of the C—H bond.

   If we were to follow our usual procedure, we would use **GO**s at the center of the $C_6$ framework of prismane to generate six MOs in each of the VAO sets $Csp^\beta$, $Csp^\gamma$, and $Cpt$. Although this approach is valid, it is more convenient to treat prismane as being composed of two triangular $C_3H_3$ fragments which we label A and B in Figure 11-3b. Separating a molecule into components is especially useful when the components possess MO systems which can be individually treated.

   Our set of VAOs for each of the $C_3H_3$ moieties of prismane is only slightly different from the set for the $N_3^+$ or cyclopropenium ions discussed in Chapter 5. By our usual procedure we can deduce the MOs for each of the $C_3$ fragments of prismane and they are listed as MO equivalence sets in Table 11-1. For the sake of argument, assume that $\alpha > \beta > \gamma$. Then the energy levels for the hybrid VAOs of these fragments can be arranged as shown in Figure 11-4. We now place **GO**s at the molecular center between identical pairs of $C_3$ fragment MOs as shown in Figure 11-5. Each such pair is thus seen to give rise to a pair of linear combinations which are either bonding or antibonding along the vertical edges of the prism (Figure 11-5a, b; c, d; e, f; g, h; etc.). These pairs are designated in the horizontal rows of Table 11-1 according to their overall bonding (b), antibonding (a), or unclear bonding (u) nature. Thus

Table 11-1. Generator Table for the $C_6$ Cluster Framework of Prismane (18 e⁻)

| Equivalence Sets | GOs | | | | | | | |
|---|---|---|---|---|---|---|---|---|
| | $s$ | $px$ | $py$ | $pz$ | $dxz$ | $dyz$ | $f3''$ | $g3''$ |
| $C_3sp^\beta a = \sigma_A(s),\ \sigma_H(s)$ | b | | | u | | | | |
| $C_3sp^\beta b = \sigma_A^*(px),\ \sigma_B^*(px),$ | | | | | | | | |
| $\qquad \sigma_A^*(py),\ \sigma_B^*(py)$ | | u | u | | a | a | | |
| $C_3sp^\gamma a = \pi_A(pz),\ \pi_B(pz)$ | b | | | u | | | | |
| $C_3sp^\gamma b = \pi_A^*(dxz),\ \pi_B^*(dxz),$ | | | | | | | | |
| $\qquad \pi_A^*(dyz),\ \sigma_B^*(dyz)$ | | u | u | | a | a | | |
| $C_3pt a = \sigma_{pA}(px),\ \sigma_{pB}(px),$ | | | | | | | | |
| $\qquad \sigma_{pA}(py),\ \sigma_{pB}(py)$ | | b | b | | u | u | | |
| $C_3pt b = \sigma_{pA}^*(f3''),\ \sigma_{pB}^*(f3'')$ | | | | | | | u | a |

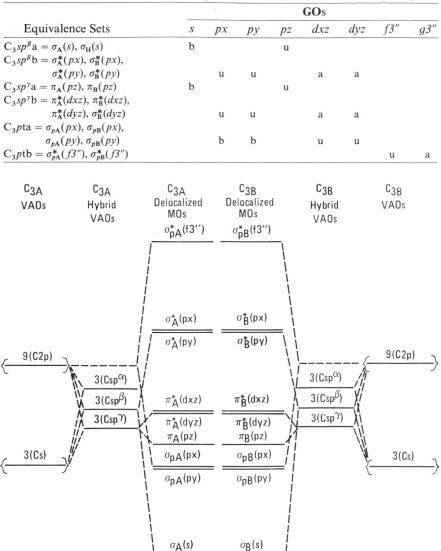

Figure 11-4. Delocalized MO energy level diagram for $C_3$ fragments of prismane. Note that the $(sp^z)$ hybrid VAOs on each fragment are assumed to form localized 2-center MOs with the hydrogens.

in Figure 11-5d, f, j, l, and r it is seen that interactions within the $C_3$ moieties as well as the interactions between the two $C_3$ fragments are antibonding and so they are identified as such in Table 11-1. Similarly, the MOs in Figure 11-5a, g, m, and o are manifestly BMOs. While the remaining MOs are clearly either bonding or antibonding between pairs of fragment MOs, their overall bonding or antibonding natures are not obvious since it is very difficult to

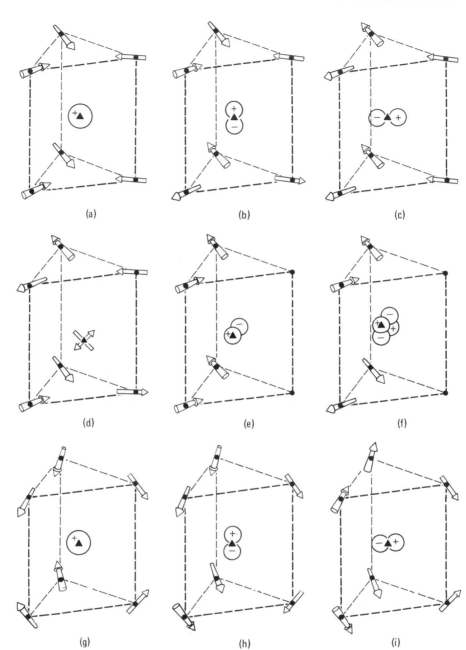

Figure 11-5. Sketches of **GO**-generated linear combinations of $C_3$ fragment MOs in prismane. The $C_3$ fragment MOs are composed of linear combinations of carbon ($sp^\beta$) [(a)–(f)], ($sp^\gamma$) [(g)–(l)], and ($2px$) [(m)–(r)] VAOs.

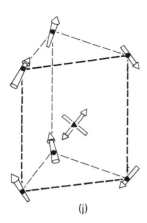

(j)

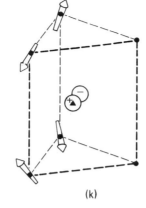

(k)

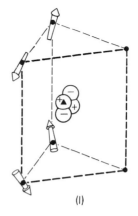

(l)

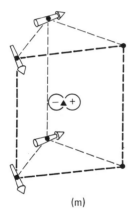

(m)

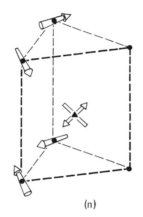

(n)

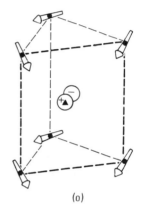

(o)

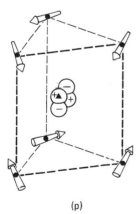

(p)

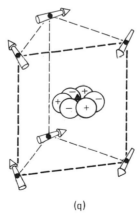

(q)

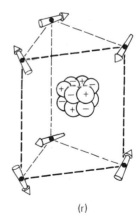

(r)

Figure 11-5 (continued)

gauge the relative importance of intra- versus inter-fragment interactions. The splittings between all the pairs of fragment MOs, as indicated by the MOs identified in the rows of Table 11-1, are depicted in Figure 11-6a. From the columns in Table 11-1 we also see that additional splittings can occur owing to identical symmetries of the MOs (i.e., MOs generated by the same **GO**) and these are shown in Figure 11-6b.

It seems appropriate that we fill the four completely bonding MOs and leave the five antibonding MOs empty. It is true, of course, that the two $s$-generated BMOs and the two $dxz$- and $dyz$-generated ABMOs will change their relative energies as a result of interaction, but they will still retain their overall bonding or antibonding character.

On the basis of our straightforward Lewis structure, we would expect to find five additional BMOs from among the nine "u"-type MOs in Table 11-1. We have no choice but to designate as BMOs a degenerate pair of "u" MOs from the $px, py$ columns of Table 11-1 and another such pair from the $dxy, dyz$ columns. (We could, of course, have designated two pairs of "u" MOs from the $px, py$ columns of Table 11-1 as BMOs, but that would presume that there are no BMOs among $dxz$-, $dyz$-generated MOs and this does not seem likely.) The remaining BMO must be either the $f3''$-generated MO or a linear combination of the $pz$-generated MOs. The choice is not obvious. Let us assume, however, that by linear combination of the $pz$-generated MOs we drive the energy of one of them well below the $f3''$-generated MO as shown in Figure 11-16b.

The occupied MOs in Figure 11-6 are generated by a set of **GO**s which includes two $s$, two $px$, two $py$, one $pz$, one $dxz$, and one $dyz$ **GO**. In the prismane geometry, the $dz^2$ **GO** generates identically with $s$ and so we can replace one of the $s$ **GO**s by a $dz^2$ **GO**. Similarly, one of the $px$ and $py$ **GO**s can be replaced by an $f1'$ and $f1''$ **GO**, respectively. This gives rise to an $sp^3d^3f^2$ hybrid **GO** set whose main lobes point toward the midpoints of the edges of a trigonal prism (i.e., $sp^2$ toward the vertical edges and $pd^3f^2$ toward the upper and lower edges). The nine localized bonds are then centered between each pair of carbon atoms with appropriate contributions from the $Csp^\beta$, $Csp^\gamma$, and $Cpt$ VAOs on each carbon. We can also rehybridize these VAOs to produce two identical hybrids pointing their major lobes over the edges of the triangular rings, and a unique hybrid directing its main lobe over a vertical edge of the prismatic $C_6$ cluster. To the extent that the C—C bonds in the rings are not equivalent to those between the two rings, the energies of the two types of localized MOs will be different. As in the delocalized bond picture, the bond order is 1.0. Note that the presence of more than one identical **GO** for generating the occupied delocalized MOs we wish to localize does not alter the bond order derived from the delocalized view. Unlike the analogous situation in $PF_5$ and $IF_7$, there is no central atom in prismane to respond to the additional **GO**s.

The molecular motions of prismane are linear combinations of the molecular motions of the $C_6$ prism and the $H_6$ prism generated by the same **GO**. The molecular motions of a prism are generated in the same manner as the

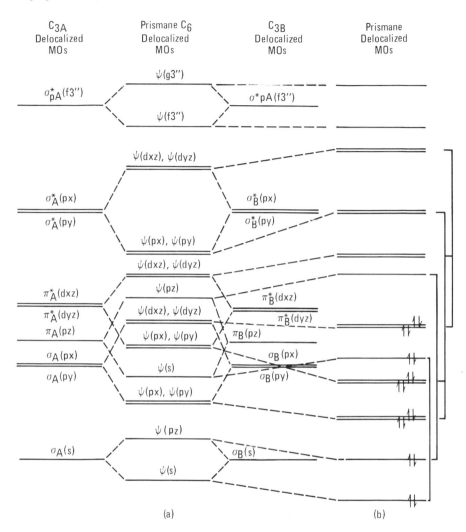

Figure 11-6. Delocalized MO energy level diagram for prismane. In (a) are shown the splittings indicated by the entries in the rows of Table 11-1 and in (b) are depicted further splittings of levels generated by the same **GO**s.

MOs were in Figure 11-5, except that $z$ in the axis system in Figure 11-3a could be pointed toward the center of gravity (i.e., the center of the molecule) with the assumption that the C—H bonds also lie along this axis.

## 11.2. C$_8$H$_8$ (Cubane)

The electron dash structure represented by Figure 11-1b indicates that there are sufficient valence electrons to achieve a bond order of 1.0 in all the links. Treating this molecule analogously to prismane we see that each of the three

Table 11-2. Generator Table for $C_8$ Framework of Cubane (24 $e^-$)

| VAO Equivalence Sets | GOs | | | | | | | | | | | |
|---|---|---|---|---|---|---|---|---|---|---|---|---|
| | $s$ | $px$ | $py$ | $pz$ | $dxy$ | $dxz$ | $dyz$ | $d(x^2-y^2)$ | $f2'$ | $f2''$ | $g4''$ | $h4''$ |
| $C_4 sp^\beta a = \sigma_A(s),\ \sigma_B(s)$ | b | | | u | | | | | | | | |
| $C_4 sp^\beta b = \sigma_A^0(px),\ \sigma_B^0(px),\ \sigma_A^0(py),\ \sigma_B^0(py)$ | | b | b | | | a | a | | | | | |
| $C_4 sp^\beta c = \sigma_A^*(dxy),\ \sigma_B^*(dxy)$ | | | | | u | | | | | a | | |
| $C_4 sp^\gamma a = \pi_A(pz),\ \pi_B(pz)$ | b | | | u | | | | | | | | |
| $C_4 sp^\gamma b = \pi_A^0(dxz),\ \pi_B^0(dxz),\ \pi_A^0(dyz),\ \pi_B^0(dyz)$ | | b | b | | | a | a | | | | | |
| $C_4 sp^\gamma c = \pi_A^*(f2''),\ \pi_B^*(f2'')$ | | | | | u | | | | | a | | |
| $C_4 pta = \sigma_{pA}^0(px),\ \sigma_{pB}^0(px),\ \sigma_{pA}^0(py),\ \sigma_{pB}^0(py)$ | | b | b | | | a | a | | | | | |
| $C_4 ptb = \sigma_{pA}(dx^2-y^2),\ \sigma_{pB}(dx^2-y^2)$ | | | | | | | | b | u | | | |
| $C_4 ptc = \sigma_{pA}^*(g4''),\ \sigma_{pB}^*(pg4'')$ | | | | | | | | | | u | | a |

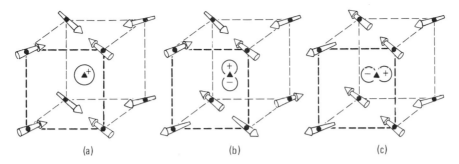

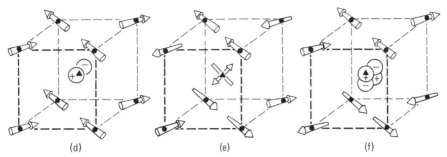

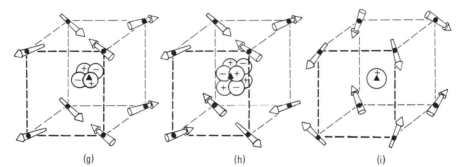

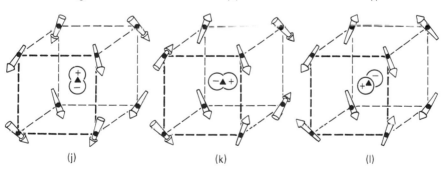

Figure 11-7. Sketches of linear combinations of C$_4$ fragment MOs generated by **GOs** in cubane. The C$_4$ fragment MOs are composed of linear combinations of carbon ($sp^\beta$) [(a)–(h)], ($sp^\gamma$) [(i)–(p)], and ($2px$) [(q)–(x)] VAOs.

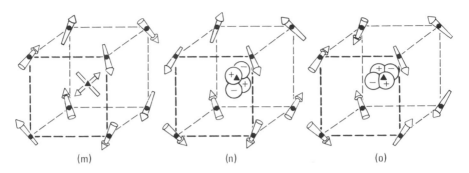

(m)                          (n)                          (o)

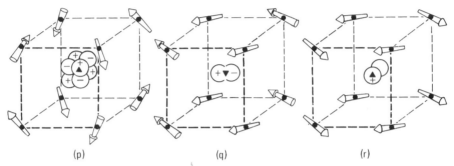

(p)                          (q)                          (r)

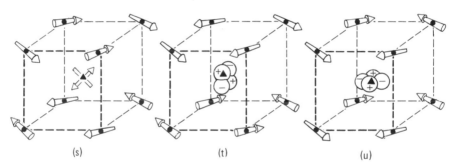

(s)                          (t)                          (u)

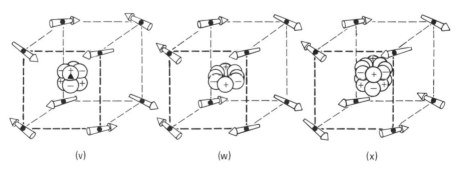

(v)                          (w)                          (x)

Figure 11-7 (continued)

VAO sets $Csp^\beta$, $Csp^\gamma$, and $C$pt in a $C_4H_4$ moiety gives rise to four MOs which have the same symmetry characteristics as the radial, $\pi$, and tangential sets, respectively, in $Te_4^{2+}$. The MOs of the frameworks of the two $C_4$ fragments A and B are grouped into eight valence orbital equivalence sets in Table 11-2. Each identical pair of $C_4$ fragment MOs can be linearly combined as shown in Figure 11-7 and they are identified in the rows of Table 11-2. These MOs are, of course, further split by the **GO** symmetry considerations implied by the entries in the columns of this table. As is also seen from this table we have nine manifestly bonding MOs and only three more are required to house the twenty-four electrons in the $C_8$ cluster. It seems reasonable to assume that linear combination of the $pz$- and $dxy$-generated "u" orbitals could give rise to two BMOs. The last MO is taken to be the $g4''$-generated "u" MO since the remaining columns would all yield manifestly antibonding MOs.

The occupied MOs are generated by a set of **GO**s which includes two $s$, three $px$, three $py$, one $pz$, one $dxy$, one $dx^2 - y^2$, and one $g4''$. Substituting a $dz^2$ **GO** for an $s$, and $f1'$, $f1''$ and $h1'$, $h1''$ for two $px$, $py$ pairs, the **GO** set becomes $sp^3d^3f^2h^2g$ which, upon hybridization, produces **GO**s whose main lobes point to the midpoints of the edges of a cube. In these regions are concentrated the electron pairs which are contained in the localized bonds. The localized bonds are made up of appropriate contributions from the $Csp^\beta$, $Csp^\gamma$, and $C$pt VAOs on each of the eight carbons. Alternatively, we can rehybridize these VAOs in a manner analogous to that used for a localized view of prismane.

## 11.3. $(\eta^5 - C_5H_5)_2Fe$

Ferrocene was the first "sandwich" compound to be synthesized. The orange sublimable crystals obtained from the reaction of $NaC_5H_5$ and $FeCl_2$ are stable in excess of $500°C$!

The electron dash structure of ferrocene shown in Figure 11-8 suggests that each ligand anion contributes six $\pi$ electrons for bonding to the metal. The iron atom can then be considered as an $Fe^{2+}$ ion, having a $3d^6$ VAO configuration. These six iron VAO electrons are shown as iron lp's in Figure

Figure 11-8. Electron dash structure for ferrocene.

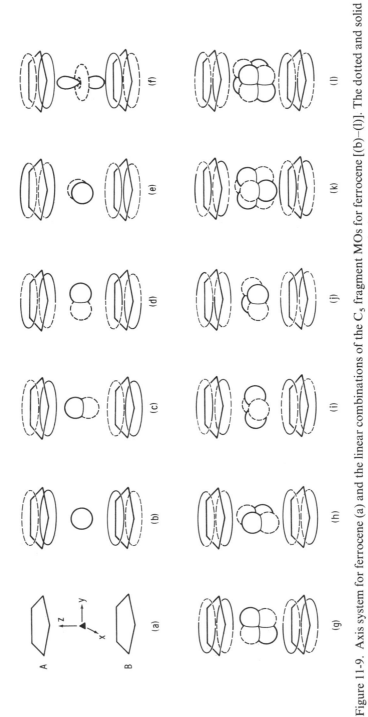

Figure 11-9.  Axis system for ferrocene (a) and the linear combinations of the $C_5$ fragment MOs for ferrocene [(b)–(l)]. The dotted and solid lines in the ligand $\pi$ MOs and **GO**s denote regions in which $\psi$ is negative and positive, respectively.

Table 11-3. Generator Table for Ferrocene (18 e)

| VAO Equivalence Sets[a] | GO's | | | | | | | | | |
|---|---|---|---|---|---|---|---|---|---|---|
| | $s(dz^2)$ | $px$ | $py$ | $pz$ | $dxy$ | $dx^2-y^2$ | $dxz$ | $dyz$ | $f2'$ | $f2''$ |
| $Feda = (Fe3dxy), (Fe3dx^2-y^2)$ | | | | | n | n | | | | |
| $Fedb = (Fe3dxz), (Fe3dyz)$ | | | | | | | n | n | | |
| $Fedc = (Fe3dz^2)$ | n | | | | | | | | | |
| $Fes = (Fe4s)$ | n | | | | | | | | | |
| $Fepa = (Fe4px), (Fe4py)$ | | n | n | | | | | | | |
| $Fepb = (Fe4pz)$ | | | | n | | | | | | |
| $Cp\pi a = \pi_A(pz), \pi_B(pz)$ | n | | | n | | | | | | |
| $Cp\pi b = \pi_A(dxy), \pi_B(dxy)$ $= \pi_A(dyz), \pi_B(dyz)$ | | n | n | | | | n | n | | |
| $Cp\pi c = \pi_A^*(f2'), \pi_B^*(f2')$ $= \pi_A^*(f2''), \pi_B^*(f2'')$ | | | | | n | n | | | n | n |

[a]The abbreviation Cp is commonly used for the cyclopentadienide ion.

11-8. It is interesting to note that the eighteen electrons around the iron (six lone pairs and twelve bond pairs) provide an electron configuration for this atom which is identical to that of the inert gas Kr at the end of the periodic row in which iron resides. Eighteen-electron metal configurations play an important role in the stability of many organometallic complexes.

Here, as in the $CoF_6^{3-}$ ion, we defer the question of orbital assignment of the transition metal lp's until later. With the aid of the axis system shown in Figure 11-9a we construct the SOs depicted in Figure 11-9b–1. The SOs are generated among the iron VAO equivalence sets and among the five delocalized $\pi$ MOs of each $C_5H_5^-$. As we saw in Chapter 7, these five $\pi$ MOs consist of one BMO, a degenerate pair of BMOs, and a degenerate pair of ABMOs. Hence these MOs group themselves into three MO equivalence sets. From the SOs depicted in Figure 11-9 Generator Table 11-3 is constructed. In this table we recognize from the symmetry of the molecule that the iron VAO

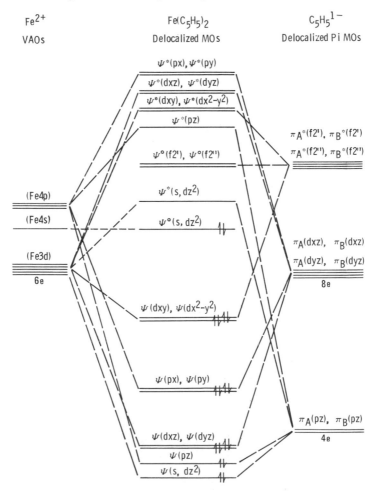

Figure 11-10. Delocalized MO energy level diagram for ferrocene.

equivalence sets Fe$a$, Fe$b$, and Fe$pa$ contain VAO partners while the VAO equivalence sets Fe$c$, Fe$s$, and Fe$pb$ contain only one VAO each.

The occupation of the MOs in ferrocene can be arrived at beginning with the assumption that seven BMOs are obtained upon linear combination of each of the SO pairs generated by the same **GO** in Table 11-3. We can expect at least one BMO from the linear combination of the SO's in the $s(dz^2)$-generated SO's, and in this MO we place two more electrons. The remaining two electrons in this diamagnetic compound must also be placed in an MO arising from this set, for to place them in the degenerate $f2'$, $f2''$-generated ABMO's would require that these electrons be unpaired. We now determine the nature of the last occupied MO. We begin by noting that the $\pi_A(pz)$, $\pi_B(pz)$ BMO's (which employ carbon $(2pz)$ VAO's) lie lower than any of the iron VAO's (i.e., the principal quantum numbers are n = 3 and 4, respectively). Since (Fe3$d$) lies below (Fe3$s$), we expect the (Fe3$d$) to interact more strongly with $\pi_A(pz)$, $\pi_B(pz)$. Thus (Fe4$s$) becomes the largely nonbonding MO in the linear combinations and it is in this orbital of largely lone pair character that we place the last pair of electrons as shown in the MO energy level diagram of Figure 11-10.

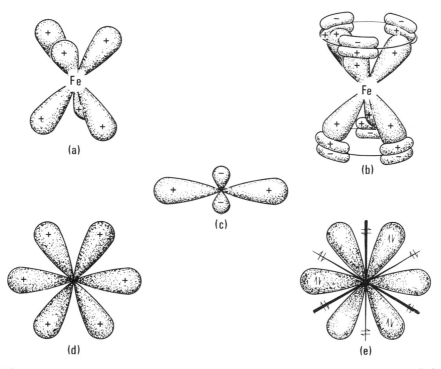

Figure 11-11. Localized orbitals in ferrocene: (a) the main lobes of a set of $sp^3d^2$ $(p^3d^3)$ hybrids; (b) the formation of localized bonds via the localized $\pi$ MOs of the C$_5$ fragments; (c) an $(sd^2)$ hybrid; (d) the orientation of the main lobes of such a hybrid set; (e) one orientation of the lone pairs in the (Fe$sd^2$) set relative to the bond pairs, as viewed down the $z$ axis of the molecule.

The Lewis structure in Figure 11-8 suggests that we should localize three bond pairs between each of the ligands and the metal. The **GO**'s which generate the six lowest BMO's in Figure 11-10 are $s$ (or $dz^2$), $px$, $py$, $pz$, $dxz$, and $dyz$. When hydridized, members of the **GO** hybrid set point their main lobes to the corners of a trigonal antiprism as shown in Figure 11-11a. The localized MO's generated by these **GO**'s consist essentially of the corresponding iron hybrid VAO's and the localized ligand MO's concentrated in those regions of the rings toward which the **GO**'s point (see Figure 11-11b). Here the ligand contributions are the localized MO's discussed for $C_5H_5{}^-$ in Chapter 7.

The **GO**s associated with the remaining three delocalized MOs in Figure 11-10 are $s$ (or $dz^2$), $dxy$, and $dx^2 - y^2$. Hybridizing these **GO**s leads to three hybrid **GO**s in the equatorial plane. A top view of one such hybrid is shown in Figure 11-11c and a top view of the main lobes of all of them is given in Figure 11-11d. The corresponding localized MOs resulting from superposing the highest three delocalized MOs will consist of the identical $(sd^2)$ [or $(d^3)$] iron hybrids and certain contributions from the ligand ions. The bonding character of $\psi(dxy)$ and $\psi(dx^2 - y^2)$ will be evenly distributed over these three localized

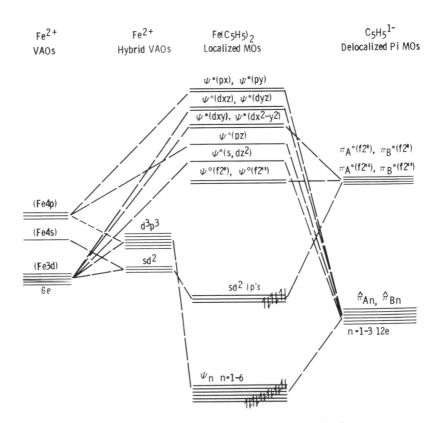

Figure 11-12. Localized MO energy level diagram for ferrocene.

Table 11-4. A Useful Summary of Procedures for Generating Pictorial Representations of Delocalized and Localized Bonding Views of a Molecule and for Visualizing its Vibrational Modes

| Delocalized View | Localized View | Normal Modes |
|---|---|---|
| 1. Decide on best electron dash structure using lowest possible formal charges and drawing resonance forms. In such forms, some lp's may be involved in a $\pi$ system. | 1. Decide on which set(s) of doubly occupied delocalized MOs to localize. | 1. Identify the motional atomic vector (AV) equivalence sets and list them in the generator table. |
| 2. Establish the molecular geometry from VSEPR theory if possible. For transition metal complexes and boron hydrides, other considerations apply. | 2. Hybridize the **GO**s which give rise to the sets(s) in 1. | 2. Using **GO**s, sketch the symmetry motions (SMs). |
| | 3. Using the hybridized **GO**s, sketch the SOs. Treat groups of inequivalent peripheral atoms separately. | 3. Fill in the rest of the generator table. (i.e., **GO**s and m entries). |
| 3. After preassigning lp's and localized bp's to ($s$) and suitably oriented ($p$) VAOs, identify all the VAO equivalence sets (hybrids may also be used) and list them in the generator table. *For groups of atoms (e.g., polygons) valence MO equivalence sets apply.* | 4. Sketch the MOs, being aware of the consequences of a central atom on which the full set of SOs (hybridized VAOs) is not generated by the hybridized **GO** set. If necessary, re-hybridize peripheral atom VAOs and rotate hybrid **GO**s to locate bp's between atoms. | 4. Take appropriate linear combinations of SMs generated by the same **GO** and sketch all the molecular motions. |
| 4. Using **GO**s located at a molecular center or centroid, sketch all the SOs, recalling that #SOs = #VAOs. Treat groups of inequivalent peripheral atoms separately. | 5. Draw the resulting MO energy level diagram and occupy the appropriate levels with electrons. | 5. Identify the normal vibrational modes. |
| 5. Fill in rest of generator table (i.e., the necessary **GO**s and the $n$, $a$ or $b$ entries). Establish partner **GO** and SO sets. | 6. Verify that the bond order per link is the same as in the delocalized view. | |
| 6. Take linear combinations of SOs generated by the same **GO** and sketch all the MOs. | | |
| 7. Draw an MO energy level diagram and occupy the appropriate levels with electrons. | | |
| 8. Deduce the bond order per link. | | |

MOs. They can be described as being largely iron lp's which are somewhat delocalized, in a bonding fashion, to the two ligand rings. These six lone pair-like lobes can, in principle, be oriented by the appropriate hybridization anywhere in this plane. On the basis of VSEPR considerations, however, one might expect that lp lobes to be staggered with respect to the antiprismatic bonding lobes in order to be maximally localized as shown in Figure 11-11e. However, influences beyond the scope of this book may dominate that would alter this conclusion. A localized energy level diagram is given in Figure 11-12.

# Summary

In this chapter the technique of treating prismatic molecules such as prismane, cubane, and ferrocene as combinations of regular polygonal moieties having valence MOs was developed. Ferrocene possesses a central atom with three "lone pairs" that we found could be housed in three metal $sd^2$ hybrids.

In Table 11-4 is the final updated version of our procedures for using **GO**s. This table is also included in Appendix V for easy reference.

PROBLEMS

1. Sketch the MOs of a $C_3$ fragment of prismane and draw the localized MO energy level diagram for this molecule.

2. Sketch the molecular motions of the $C_6$ cluster of prismane assuming that the $z$ axis in Figure 11-3a points to the center of gravity of the $C_6$ cluster. Hint: demonstrate for $Te_4^{2+}$ that it is possible to take linear combinations of SMs of like symmetry of two identical halves of a molecule to obtain the molecular motions. As a simple example, treat $Te_4^{2+}$ as a pair of diatomic molecules and take linear combinations of their SMs generated by the same **GO**.

3. Using sketches, show that a set of pure $(Csp^3)$ orbitals could be used to make localized bonds in prismane and in cubane. In these molecules, these hybrids can not be pure $sp^3$. Comment on the latter statement and qualitatively compare more realistic hybrid sets in these two molecules. Comment on the possibility of direct head-on overlap of adjacent carbon hybrids along the edges of these molecules as opposed to angular overlaps.

4. Sketch delocalized and localized MO energy level diagrams for cubane.

5. What is the bond order in ferrocene? Compare your answer with the one implied by its Lewis structure.

6. Generate the delocalized and localized views of $Te_8^{8+}$ assuming that it has a cubic molecular structure.

7. The $XeF_8^{2-}$ ion is known and in $(NO_2)_2[XeF_8]$ this ion possesses a square antiprismatic array of fluorines with a Xe atom at the center. Generate the delocalized and localized views.

8. On the basis of the delocalized MO diagram for $[Cr(C_6H_6)_2]^{1+}$ give a reason why this ion is reasonably stable in spite of its non-conformance to the eighteen-electron rule.

9. The complex $[U(C_8H_8)_2]^0$ with two unpaired electrons can synthesized by simply heating finely divided uranium metal with cyclooctatetraene. (a) What is the formal oxidation state of the metal? (b) Develop the delocalized MO picture assuming that the uranium $(5f)$, $(6d)$, $(7s)$, and $(7p)$ VAOs all participate in MO formation with the ligand orbitals.

10. The $Rh_6C(CO)_{15}{}^{2-}$ ion consists of a trigonal prismatic array of metal atoms with a carbon atom in the center. Each rhodium bears a terminal CO group and three CO groups which bridge an edge as indicated in the partial structure below. Each metal atom can be considered to be bound by two electrons from a terminal CO. Each bridging CO is bound to a metal by one carbon electron and one metal electron. The Rh—C links can be assumed to utilize a set of roughly tetrahedrally oriented hybrids obtained by mixing the $(5s)$, $(4dxy)$, $(4dyz)$, and $(4dxy)$ VAOs on rhodium. Develop the delocalized bonding view for the $Rh_6C$ cluster system, assuming that all the metal atoms are close enough to interact with one another.

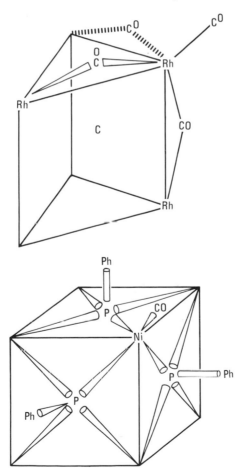

11. The cubic cluster $Ni_8(PC_6H_5)_6(CO)_8$ has the structure indicated in partial form above. Each nickel binds to a phosphorus using one phosphorus electron and one nickel electron and the terminal CO binds with two carbon electrons. Each $P_3NiC$ moiety is assumed to employ a set of tetrahedrally deployed $sd^3$ metal hybrids obtained by mixing the nickel $(4s)$, $(3dxy)$, $(3dxz)$ and $(3dyz)$ VAOs. Develop a delocalized bonding view of the $Ni_8$ cluster.

# Symmetry, Molecular Orbitals, and Generator Orbitals

Molecules, such as $H_2$, intuitively appear to be less symmetrical than atoms. Thus a hydrogen atom has a spherical nucleus and a spherical electron density. A hydrogen molecule possesses two nuclei and its electron density resembles a sausage. The peculiar shapes of AOs are a consequence of the spherical symmetry of atoms. The physical scientist defines symmetry in terms of symmetry elements. These elements apply to any object, and they represent physical manipulations of or "operations" on the object which do not allow an observer to detect that any operation occurred if he/she had been temporarily blindfolded. For example, consider the equilateral triangle (Figure 1) which has three identical vertices. One of its symmetry elements is a three-fold axis $Rot_3$ perpendicular to its plane. If the triangle is rotated around this axis once, twice, or three times by 120° (in either a clockwise or counterclockwise direction), a temporarily blindfolded observer would not realize what had occurred since all the points of the triangle are identical. We say that the operation associated with this symmetry element leaves the triangle "invariant". There are also three two-fold rotation axes (Figure 1(b)) about which rotations by 180° leave the triangle invariant. Similarly there is a reflection plane (Figure 1(c)) through which any point in the top half of our triangular object reflected directly across this plane to a point equidistant below the plane leaves the triangle invariant (Figure 1d). There are also three reflection planes perpendicular to the plane of the triangle. In addition, there is another symmetry element called the "identity" element whose associated operation leaves the object untouched and hence invariant. (There is one more symmetry element, namely an improper axis of rotation which need not concern us here.)

A spherical object (such as an atom) has all of the symmetry elements of a triangle plus many more. Although a sphere (like any other object) has only one identity operation, it possesses an infinite number of $n$-fold rotation axes

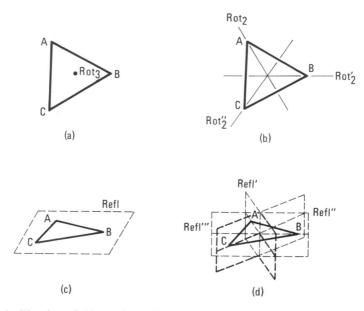

Figure 1. The three-fold rotation axis (a), two-fold rotation axes (b), and planes of reflection (c), (d) of an equilateral triangle having three equivalent vertices.

where $n$ can be any number from one to infinity. It also contains an infinite number of reflection planes. Hence a sphere is the most symmetrical geometrical form we know of. A triangle, or any object which is not a sphere, must therefore possess lower symmetry, and this conclusion is reflected by the fact that the symmetry elements of nonspherical objects in every case constitute a subset of the symmetry elements of a sphere. Because of this fact, the shapes of canonical AOs (i.e., their nodal and lobal patterns) can be related to the symmetry of any molecule and hence also can be related to the shapes of the molecule's orbitals. In other words, molecular orbitals have nodal and lobal patterns that are analogous to those of AOs.

The question is, how do we construct or generate MOs for a given molecule? We must find an appropriate method which compares molecular orbitals with AOs while taking into account the symmetry of the molecule we are considering. One such method is embodied in group theory which is generally taught in a graduate level course. A simpler (but nonsimplistic) method is the one developed in this book, which utilizes atomic orbitals as pictorial generators of molecular orbitals.

# A Kit for Visualizing MOs in Three Dimensions

Atomic orbitals and the spatial orientations they assume in symmetry orbitals in various molecular geometries can be visualized using a set of simple three-dimensional models. A kit[a] for this purpose can be easily constructed from Styrofoam orbitals and a wooden support as illustrated in Figures 1 and 2, respectively. In Figure 3 is an example of how the apparatus can be used to view the $px$- and $py$-generated SOs of the (Cl$3pz$) VAOs in PCl)$_3$. The kit can be made in a range of sizes. A convenient size for the base is about $6'' \times 6'' \times \frac{3}{4}''$ with orbitals constructed from $1''$ and $1\frac{1}{4}''$ Styrofoam balls and $9'' \times 3/16''$ dowels. Tempera paints can be used to denote the different signs of the orbital wave functions.

[a] J.G. Verkade and K. Ruedenberg "Mechanical System for Visually Determining Permitted Electron Distributions within Molecules", U.S. Patent 3,967,388, July 6, 1976.

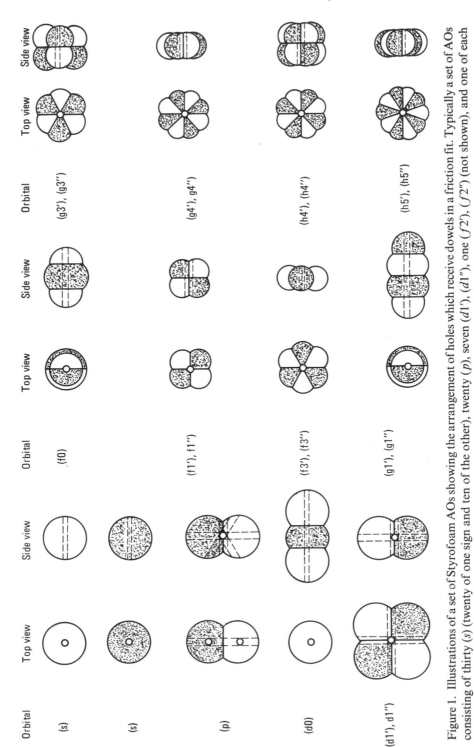

Figure 1.  Illustrations of a set of Styrofoam AOs showing the arrangement of holes which receive dowels in a friction fit. Typically a set of AOs consisting of thirty (s) (twenty of one sign and ten of the other), twenty (p), seven (d1'), (d1''), one (f2'), (f2'') (not shown), and one of each of the remaining AOs shown is useful.

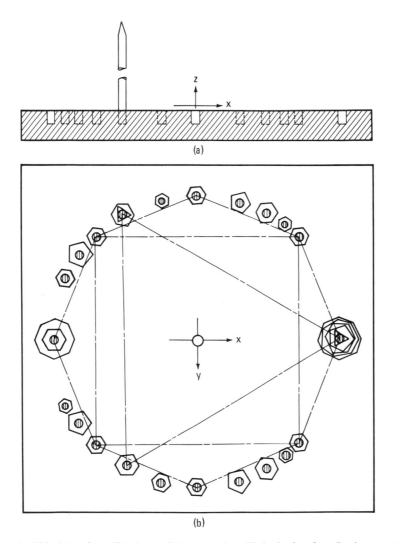

Figure 2. Side (a) and top (b) views of the base plate. Holes in the plate (in the top view) are located at the vertices of regular polygons and receive friction-fitting dowels which may be pointed for easier insertion into the orbitals.

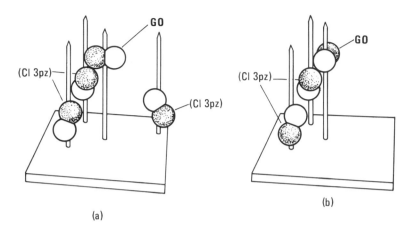

Figure 3.  Arrangement of *px*- (a) and *py*-generated (b) SOs of the chlorine ($3pz$) VAOs in $PCl_3$.

# Hybrid GOs and Non-Cooperating Central Atoms

Here we show that there must be a small antibonding contribution to the localized MOs of FHF$^-$ shown in Figure 4-6b. Thus it will be shown that in the expressions below for these MOs, the coefficient $\alpha$ is smaller than $\beta$ and of opposite sign.

$$\hat{\sigma}_A = \alpha(F_B 2pz) + \beta(F_A 2pz) + \gamma(H1s) \tag{1}$$

$$\hat{\sigma}_B = \beta(F_B 2pz) + \alpha(F_A 2pz) + \gamma(H1s). \tag{2}$$

For normalization we require that

$$
\begin{aligned}
1 = \langle \hat{\sigma}_A | \hat{\sigma}_A \rangle = \alpha^2 + \beta^2 &+ 2\alpha\beta \langle (F_A 2pz) | (F_B 2pz) \rangle \\
&+ 2\alpha\gamma \langle (F_B 2pz) | (H1s) \rangle \\
&+ 2\beta\gamma \langle (F_A 2pz) | (H1s) \rangle,
\end{aligned} \tag{3}
$$

which can be simplified to

$$1 = \alpha^2 + \beta^2 + \gamma^2 + 2\gamma(\alpha + \beta)\langle (F2pz) | (H1s) \rangle, \tag{4}$$

if we assume that the overlap between $(F_A 2pz)$ and $(F_B 2pz)$ is zero and that the values of the wave functions $(F_A 2pz)$ and $(F_B 2pz)$ are the same. The expression in Equation 4 is identical for $\langle \hat{\sigma}_B | \hat{\sigma}_B \rangle$. For orthogonality we require that

$$
\begin{aligned}
0 = \langle \hat{\sigma}_A | \hat{\sigma}_B \rangle = (\alpha^2 + \beta^2) &\langle (F_A 2pz) | (F_B 2pz) \rangle + \gamma^2 + 2\alpha\beta \\
&+ \gamma(\alpha + \beta)\langle (F_A 2pz) | (H1s) \rangle \\
&+ \gamma(\alpha + \beta)\langle (F_B 2pz) | (H1s) \rangle
\end{aligned} \tag{5}
$$

which can be simplified to

$$0 = \gamma^2 + 2\alpha\beta + 2\gamma(\alpha + \beta)\langle (F2pz) | (H1s) \rangle, \tag{6}$$

if we invoke the same assumptions used in writing Equation 4. In order to simplify our task still further, we make use of the so-called "zero neighbor overlap approximation," which says that the overlap between neighboring atoms is in fact usually much less than unity because the overlap integral is non-zero only in regions of space where *both* VAOs are appreciably non-zero. Hence Equations 4 and 6 can be reduced to Equations 7 and 8, respectively,

$$1 = \alpha^2 + \beta^2 + \gamma^2, \tag{7}$$

$$0 = \gamma^2 + 2\alpha\beta, \tag{8}$$

for qualitative purposes. It is reasonable to assume also that ($F_B 2pz$) and (H1$s$) in Equation 1 are fully called in, with the consequence that the coefficients of these functions in Equation 1 are the same (i.e., both VAOs are fully engaged in this MO). By substituting $\alpha = \gamma$ into Equation 8, we obtain the result that $0 = \alpha(\alpha + 2\beta)$ for which the only reasonable solution is that $-\frac{1}{2}\alpha = \beta$. By substituting $-\frac{1}{2}\alpha = \beta$ ($= -\frac{1}{2}\gamma$) into Equation 7, we find that the normalized values of these coefficients are $\alpha = 2/3^{1/2} = \gamma$ and $\beta = -3^{-1/2}$. Thus we conclude that there are indeed negative contributions of one of the ($F2pz$) VAOs in each of the MOs $\hat{\sigma}_A$ and $\hat{\sigma}_B$, which are only half as large as the contributions of the other ($F2pz$) VAO. Moreover, these different ($F2pz$) contributions are necessary to preserve orthogonality between $\hat{\sigma}_A$ and $\hat{\sigma}_B$. Thus in Equation 8, $\beta$ could not possibly be zero, for then $\gamma$ would have to be zero too and there would be *no* hydrogen (1$s$) contribution.

    In FXeF (Chapter 4.2) there is a situation similar to that described above for FHF$^-$ except that there is a xenon ($p$) VAO which is the only central atom VAO called in by the $sp_l$ and $sp_r$ **GO**s. By applying analogous reasoning to expressions for the MOs depicted in Figure 4-11, it can be shown that there are again smaller contributions from one of the ($F2pz$) VAOs but this time these contributions are not opposite in sign to that of the larger ($F2pz$) coefficient.

# Calculating the Angles for Conal Nodes

In Table 1 are set out the dependences of hydrogen-like AOs on the angle $\theta$ (see Figure 2-6). If we wish to calculate this angle for, say, ($f1'$) or ($f1''$), we set the expression equal to zero because we desire the location where the wave function vanishes:

$$(f1')_\theta = (f1')_\theta = \frac{\sqrt{42}}{8} \sin\theta(5\cos^2\theta - 1) = 0. \tag{1}$$

Under these circumstances the normalization coefficient $\sqrt{42}/8$ drops out. One of the roots is $\sin\theta = 0$, which reflects the fact that there is a trivial root for which $\theta = 0$. The remainder of Equation 1 reduces to $\cos\theta = \pm\sqrt{1/5}$, from which we find that there are two angular nodes, one at $63°26'$ and the other at $180°\ 63°26'$.

Table 1. Hydrogen-Like AO Functions having Angular Nodes

| AO | Wave Function |
|---|---|
| $(p)$ | $(p0)_\theta = \dfrac{\sqrt{6}}{2}\cos\theta$ |
| $(d)$ | $(d0)_\theta = \dfrac{\sqrt{10}}{4}(3\cos^2\theta - 1)$ |
| | $(d1')_\theta = (d1'')_\theta = \dfrac{\sqrt{15}}{2}\sin\theta\cos\theta$ |
| $(f)$ | $(f0)_\theta = \dfrac{3\sqrt{14}}{4}\cos\theta(\cos^2\theta - 1)$ |
| | $(f1')_\theta = (f1'')_\theta = \dfrac{\sqrt{42}}{8}\sin\theta(5\cos^2\theta - 1)$ |
| | $(f2')_\theta = (f2'')_\theta = \dfrac{\sqrt{105}}{4}\sin^2\theta\cos\theta$ |
| $(g)$ | $(g0)_\theta = \dfrac{9\sqrt{2}}{16}\left(\dfrac{35}{3}\cos^4\theta - 10\cos^2\theta + 1\right)$ |
| | $(g1')_\theta = (g1'')_\theta = \dfrac{9\sqrt{10}}{8}\sin\theta\cos\theta\left(\dfrac{7}{3}\cos^2\theta - 1\right)$ |
| | $(g2')_\theta = (g2'')_\theta = \dfrac{3\sqrt{5}}{8}\sin^2\theta(7\cos^2\theta - 1)$ |
| | $(g3')_\theta = (g3'')_\theta = \dfrac{3\sqrt{70}}{8}\sin^3\theta\cos\theta$ |
| $(h)$ | $(h0)_\theta = \dfrac{15\sqrt{22}}{16}\cos\theta\left(\cos^4\theta - \dfrac{14}{3}\cos^2\theta + 1\right)$ |
| | $(h1')_\theta = (h1'')_\theta = \dfrac{\sqrt{165}}{16}\sin\theta(21\cos^4\theta - 14\cos^2\theta + 1)$ |
| | $(h2')_\theta = (h2'')_\theta = \dfrac{\sqrt{1155}}{8}\sin^2\theta\cos\theta(3\cos^2\theta - 1)$ |
| | $(h3')_\theta = (h3'')_\theta = \dfrac{\sqrt{770}}{32}\sin^3\theta(9\cos^2\theta - 1)$ |
| | $(h4')_\theta = (h4'')_\theta = \dfrac{3\sqrt{385}}{16}\sin^4\theta\cos\theta$ |

# A Summary of Uses for Generator Orbitals

In Table 1 is summarized the applications of **GO**s discussed in this book.

Table 1. A Useful Summary of Procedures for Generating Pictorial Representations of Delocalized and Localized Bonding Views of a Molecule and for Visualizing its Vibrational Modes

| Delocalized View | Localized View | Normal Modes |
|---|---|---|
| 1. Decide on best electron dash structure using lowest possible formal charges and drawing resonance forms. In such forms, some lp's may be involved in a $\pi$ system. | 1. Decide on which set(s) of doubly occupied delocalized MOs to localize.<br><br>2. Hybridize the **GO**s which give rise to the sets(s) in 1.<br><br>3. Using the hybridized **GO**s, sketch the SOs. Treat groups of inequivalent peripheral atoms separately. | 1. Identify the motional atomic vector (AV) equivalence sets and list them in the generator table.<br><br>2. Using **GO**s, sketch the symmetry motions (SMs).<br><br>3. Fill in the rest of the generator table. (i.e., **GO**s and m entries). |
| 2. Establish the molecular geometry from VSEPR theory if possible. For transition metal complexes and boron hydrides, other considerations apply.<br><br>3. After preassigning lp's and localized bp's to (s) and suitably oriented (p) VAOs, identify all the VAO equivalence sets (hybrids may also be used) and list them in the generator table. For groups of atoms (e.g., polygons) valence MO equivalence sets apply. | 4. Sketch the MOs, being aware of the consequences of a central atom on which the full set of SOs (hybridized VAOs) is not generated by the hybridized **GO** set. If necessary, re-hybridize peripheral atom VAOs and rotate hybrid **GO**s to locate bp's between atoms. | 4. Take appropriate linear combinations of SMs generated by the same **GO** and sketch all the molecular motions.<br><br>5. Identify the normal vibrational modes. |

Table 1 (continued)

| Delocalized View | Localized View | Normal Modes |
|---|---|---|
| 4.  Using **GO**s located at a molecular center or centroid, sketch all the SOs, recalling that #SOs = #VAOs. Treat groups of inequivalent peripheral atoms separately. | 5.  Draw the resulting MO energy level diagram and occupy the appropriate levels with electrons. | |
| | 6.  Verify that the bond order per link is the same as in the delocalized view. | |
| 5.  Fill in rest of generator table (i.e., the necessary **GO**s and the *n*, *a* or *b* entries). Establish partner **GO** and SO sets. | | |
| 6.  Take linear combinations of SOs generated by the same **GO** and sketch all the MOs. | | |
| 7.  Draw an MO energy level diagram and occupy the appropriate levels with electrons. | | |
| 8.  Deduce the bond order per link. | | |

# Index